JANE'S DELICIOUS HERBS

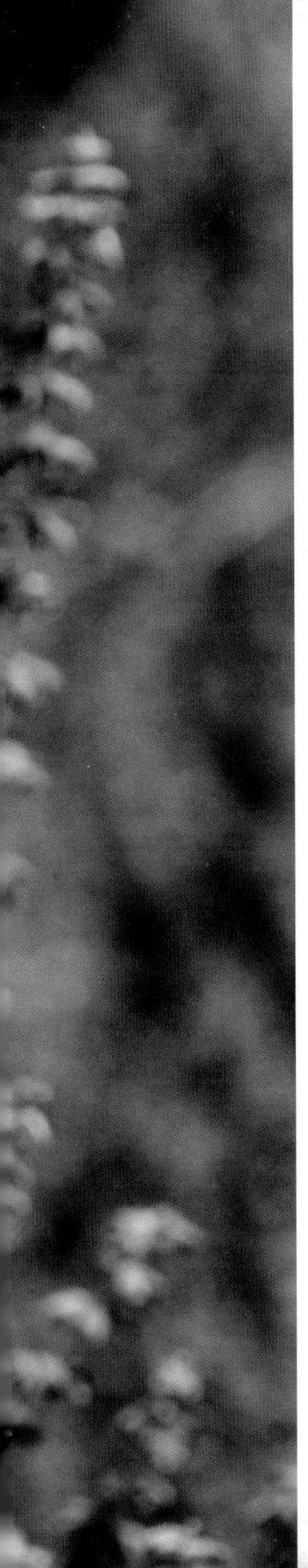

JANE'S DELICIOUS HERBS

GROWING AND USING HERBS IN SOUTH AFRICA

JANE GRIFFITHS

Photography by Keith Knowlton and Jane Griffiths

SUNBIRD PUBLISHERS

'Let food be thy medicine.'
Hippocrates

For my Garden Angels

Thanks to:
Margaret Roberts for leading the way;
Ceri, Michelle and Marius for being The Dream Team;
Ali, Apple and Noodles for all the licks on the forehead;
and, as always, to Keith for enjoying my experiments.

CONTENTS

Introduction	8
Growing Herbs	10
A–Z of Herbs	40
The Herbal Pharmacy	176
Herbal Recipes for Wellbeing, Healing and Happiness	198
A Quick Guide to Healing Herbs	260
Bibliography	291
Acknowledgements	291
Index	292

INTRODUCTION

I am the daughter of a pharmacist and as a child I believed there was a pill for every ailment. Pharmacists actually made pills then, and there was something magical about the pestle and mortar and the smell of the ingredients. Playing hide and seek amongst the cavernous shelves, while Dad measured out healing concoctions, I fully believed there was a cure for everything.

Unluckily I was wrong. Science, for all its progress, does not have all the answers. Realising that modern medicine is as fallible as its human practitioners, I began exploring a more holistic approach to healing. When I started growing my own vegetables and herbs, it was a natural step to experiment with making unguents, lotions and potions. Mom's old double boiler was hauled out from the back of the cupboard and put to good use melting beeswax and shea butter. Abundant herb harvests were chopped, macerated, dried, steeped and spread. Making everything from delicious body creams to healing tinctures, I began following in my father's footsteps.

I started by growing and using herbs that were useful for my ailments. The animals came next: anti-flea herbs for bedding and sprays to repel flies. Family and friends followed in line at my garden apothecary: comfrey poultices for sprained ankles, Dr Feelgood tea for the blues, and rose pelargonium hand cream for birthday presents.

And so, as I process herbs using my father's old pestle and mortar, I feel connected to a long line of healers who have come before me. There is something deeply satisfying about going into my garden, picking a selection of herbs and flowers, and brewing them into tea to make someone feel better. And to use homemade hand cream where I know every single ingredient and its origin. Or simply to add freshly picked parsley to scrambled eggs.

Through the years I have grown a wide variety of herbs, many of which can be grown on a windowsill or in a small garden. Most herbs are easy to grow and many are perennial and relatively hassle free. Whether you allocate a separate space for them or grow them in amongst your existing vegetable or flower garden, herbs add colour, texture and fragrance to your garden. They also have a multitude of uses – with new ones being discovered every day.

This book is not meant to be a panacea for all ills. It is a guide for those who want to move towards living a healthier lifestyle. *Jane's Delicious Herbs* is about growing and using healing plants. From adding flavour to our food, to treating us medicinally or simply making us feel good, herbs enrich our lives.

As I process herbs using my father's old pestle and mortar, I feel connected to a long line of healers who have come before me

GROWING HERBS

When I first started growing herbs, I designed a beautiful semi-circular herb garden, with radiating paths and brick edging. For the first couple of years I grew only herbs in there, but one winter I needed more space for my large cauliflowers and stole a corner for them. It was the beginning of the end of my neat herb garden, as every year more vegetables snuck into their space. I now grow herbs in amongst my vegetables as well as in beds and dedicated containers closer to my kitchen.

Herb garden design can roughly be divided into formal and informal styles.

In a formal herb garden, hard landscaping and the plants themselves, trimmed into hedges, are used to define geometrical shapes and symmetrical patterns. A formal herb garden requires maintenance, patience and plenty of planning.

An informal herb garden doesn't have as many permanent structures as a formal one and is more flowing, with curved lines. Rustic elements, such as stone or log edgings and bark pathways, create a natural feel. Plants grow naturally and sprawl out of place. Although an informal herb garden does require planning and maintenance, it is not as much work as one planted in a formal style.

Design

There are many herb garden design styles, ranging from formal French layouts to rambling cottage gardens. The design you choose should match your taste and requirements. The size of your herb garden depends on the number and type of herbs you will be planting and using. If you are growing a few herbs for your kitchen, then a couple of pots near your kitchen door will do. A larger herb garden is required if you plan to experiment a bit more.

Although a herb garden is a practical addition, it is also an opportunity to design something of beauty that will enhance your garden. It then becomes the place where you go to pick some basil for the pasta – and the place where you go to meditate and soothe your soul.

Knot gardens, dating back to the mid-fifteenth century, were originally designed to imitate elaborate medieval embroideries. This formal design makes use of low, permanent hedges of clipped herbs such as rosemary and lavender, laid in intricate patterns, which define the shapes of the beds. The beds are filled with herbs and flowers of varying heights and colours, following a rhythmic and interwoven design. The large knot gardens of the French and English palatial grounds are designed to be viewed from above, as if looking down on a carpet. The knot garden style can be adapted to home gardens, using less intricate patterns. To maintain the shapes, the herb hedges need to be trimmed regularly. Choose low-growing and compact herbs, rather than sprawling ones that will grow out of their allotted boundaries.

A **herb wheel** is a circular layout with the pathways forming the 'spokes' of the wheel, both dividing the beds and leading towards a central feature. Used for centuries, this herb garden design is a practical and attractive choice, suitable for small and large gardens. Each bed can be planted with herbs having similar requirements. An effective method is to create or use an existing small hill or mound. Place the centre of the wheel at the top and plant more drought tolerant and hardy herbs on the upper levels, with the water-needy ones further down the slope.

A **spiral garden** is a circular herb garden with an outer rim path and a second path that spirals towards the centre. This reduces the amount of space given over to pathways. In larger gardens, the spiral can have smaller connecting pathways for ease of access. As with the herb wheel, the centre of the spiral can be placed at the top of a mound.

The ancient **labyrinth** design has been used for more than 4000 years as a walking meditation to connect with nature and one's inner self. It has one path, looping its way to the centre and back out again. This is a magical feature to create in a large garden, using herbs to edge the pathways. A single herb, such as lavender, can be planted, or the space between the pathways can be filled with a variety of herbs, adding a further dimension of colour and scent to the meditative experience.

Potager means 'kitchen garden' in French. And the French always do things with style and grace. Instead of a simple utilitarian herb garden, a potager herb garden combines the practicality of a food garden with ornamental elements, such as statues, archways and water features. This can be done in a formal style, with geometric beds, or informally, with rambling plants and curved lines.

ABOVE: Knot gardens, dating back to the mid-fifteenth century, were originally designed to imitate elaborate medieval embroideries.

BELOW: A herb wheel is a practical and attractive design choice, suitable for small and large gardens.

13 GROWING HERBS

A **cottage-style herb garden** is the most informal of all. This natural style makes use of dense plantings, mixed beds and a wide variety of flowers and plants to create a slightly chaotic and rambling look. Meandering pathways or simple stepping stones between the plants add to the unstructured feel. They need less maintenance than formal herb gardens because they are supposed to be on the wild side. This style can be used in gardens of various sizes and is a good choice for gardeners with limited time.

A **container herb garden** is the ideal choice for small gardens. Many herbs grow well in containers and, even with my relatively large space, I still grow plenty of herbs in containers near my kitchen, simply because it is much easier to nip out the back door to harvest some quick herbs for a meal. There is more on container herb gardens on page 32.

A **stepping stone herb garden** is a patchwork style where squares or circles of stone are laid out in an informal or formal pattern (chequerboard works well) and the gaps in between are planted with herbs. This is a good option if you have a section of paving or lawn that you want to convert into a herb garden.

A chequerboard design is a good option if you have a section of paving that you want to convert into a herb garden

SITE

The ideal spot for a herb garden is north facing, near your kitchen, with at least five hours of sunshine a day. But if this isn't possible in your garden, don't despair: you will still be able to grow some herbs as they have varying requirements. Most herbs prefer a sheltered environment with well-drained soil.

If you have a choice between a spot with a good aspect or good soil, go for the better aspect. You can always improve soil but you can't move the sun. Assess where the shade in your garden falls and how this will change during the year. A good way of doing this is to watch the path of the full moon. It is roughly where the sun will be in six months' time. So, if you are planning your herb garden in the middle of winter, the shadows from the full moon will show you where the mid-summer sun's shadows will be.

Most of the flavour and beneficial qualities of Mediterranean herbs such as rosemary, thyme and oregano are found in their oils. These are produced in greater abundance when the herbs are grown in full sun for six to eight hours a day. However, herbs such as rocket, coriander and even basil are quite happy with far less sun than that.

Many adverse conditions can be improved with simple changes. If you don't have sufficient sunlight, cut lower branches off overhanging trees to let more light in. Paint any surrounding walls white and use pale colours on your pathways to reflect light into your herb garden. To shelter a windy area, observe where most of the wind comes from and block it with a simple split-pole fence.

OPPOSITE: Many labyrinth designs combine visual symmetry with a surprising length of pathway in a relatively small area. The lavender labyrinth at Kew Gardens is a good example of this.

Although herbs will flourish if grown in the right conditions, most are forgiving plants and even if they don't receive the optimum amount of sunshine they will still grow. Herbs grown in less sunlight tend to become straggly and leggy, which gives you good reason to harvest more often – trimming encourages bushier growth. They will need less water than those grown in sunnier areas so watch out you don't waterlog them. They will also be more vulnerable to disease and bugs, which means a bit more vigilance is necessary.

Because herbs have varying requirements, you might find you are better off choosing two or three different areas around the garden for planting herbs. There can be varied microclimates within even a small garden. A sheltered bed surrounded by paving next to a north-facing wall will be much warmer and drier than an area next to a damp, south-facing wall. An area near a tap will usually be moist and is a good spot for herbs that don't mind having damp feet. Herbs that like warmer, dry conditions will be far more suited to a hot, north-facing bed.

Many herbs flourish in containers and this is an ideal choice for small gardens, balconies and patios (see page 32 for more on growing herbs in containers). Or you can plant herbs amongst your flowers in existing beds. Instead of your beds just being decorative, they will have a useful herb or two included.

Soil

Organic herb gardening all begins with the soil. If you have a problem with your plants it is likely that you have unhealthy soil. Building up healthy, nutrient-rich soil means your plants will be stronger with more resistance to pests.

Healthy soil is full of humus – broken down organic matter, which is the 'life force' of the soil. It provides a home for billions of organisms, such as fungi, bacteria, algae, insects and worms. In one teaspoon of healthy soil there are more than six billion organisms. Without these, plants cannot grow. There are endless complex relationships going on underground: soil cleansing and enrichment, nutrient exchanges and many others that we are only just discovering. Earthworms, for example, leave the soil eight times richer after they have ingested it. If our soil is full of these billions of beneficial organisms, it means nature will do much of our work for us.

Organic herb gardens need as much humus in the soil as possible for a number of other reasons:
- Humus acts as a sponge, absorbing and storing water underground.
- Chemically, humus has numerous active surfaces, which hook nutrients. This makes many more nutrients available to plants.
- Humus improves the physical structure of soil, making it moist, crumbly and aerated, providing the ideal home for beneficial bacteria and other organisms.

Organic herb gardening begins with the soil. Building up healthy, nutrient-rich soil means your plants will be stronger with more resistance to disease

OPPOSITE: Pathways and arches create a pleasing effect in larger herb gardens.

ABOVE: Most herbs require well-drained, fertile soil and it is worth enriching your soil before beginning to plant.

The first step to creating healthy soil is to enrich it well with plenty of organic matter. Most herbs require well-drained, fertile soil and it is worth enriching your soil before beginning to plant. Once you have decided on your design and marked where the beds will be, dig out the topsoil layer and set it aside. Then dig down to a depth of about 30 cm. Use a fork to loosen and break up the layer of soil below. Add a 5–8 cm layer of compost and well-rotted manure. Add the soil back in, replacing the topsoil last. After this initial preparation, you will never need to dig your soil again.

As we harvest our herbs, we remove nutrients the plants have absorbed from the soil and, to keep our soil healthy, these need to be replaced. Do this by regularly adding organic matter (compost, manure and mulch) to the surface of the beds – but don't dig it in. Nature is designed to incorporate material that has fallen on the surface into the bottom layers. Earthworms come up at night and pull it down into the soil and in no time it will be converted into humus for your plants' roots.

No-dig gardening

Early on in my gardening journey I discovered both the joys and benefits of no-dig gardening. In many gardens it is a tradition to regularly dig up and turn over the soil, adding compost and manure, breaking up clods and aerating the soil. But this continual digging is harmful to the soil. Every time you dig up the soil you are destroying billions of beneficial organisms and breaking up their homes. While they are busy re-establishing their balance they are not getting on and doing all their essential jobs in the soil.

Digging up and turning over the soil is not only harmful to all the beneficial organisms, but it also causes moisture and nutrient loss. And, as anybody who has ever turned over the soil knows, it unearths weed seeds that have been waiting for you to expose them to the air so they can germinate and flourish. And finally, who enjoys digging anyway? No matter how much we approach it as a meditation and not a chore, it is a pain in the back. The only time I dig deep into my garden is to remove a deep-rooted weed or perennial, to harvest roots of a plant or when preparing a new bed. If you continually add organic matter to the surface of the soil, you will maintain high levels of humus.

The reason why many people dig their beds is to loosen and aerate compacted soil. And the main cause of compacted soil is our body weight pressing down on it. So the first rule of no-dig gardening is to never, ever stand on the soil. A second rule, therefore, is to make your garden beds just wide enough for you to reach the middle comfortably from the path. They should also be just long enough to walk around comfortably, so you aren't tempted to 'just this once' stand in the middle. If your existing beds are bigger than this, place stepping stones where necessary. It is a good idea to create permanent pathways around your beds. These are the safe zones where you can walk. To keep the humus-rich soil inside the beds, instead of spilling over into the paths, create an edging around your beds. This can be made from logs or stones, herb hedges, recycled bricks, planks, bottles or anything else that suits your garden design. If you are working in a bed, constructing a tripod for example, then place a plank over the soil to distribute your weight evenly.

Selecting your plants

Your herb garden begins with a choice of using seeds, seedlings or young plants. There are many specialist nurseries that supply all the well-known herbs as well as propagating a variety of interesting and obscure ones. When buying a more unusual herb, try to find out as much as you can from the nursery about its growing conditions. It is depressing to buy a plant, bring it into your garden and then over- or under-water it until it dies.

When buying herbs, try to find out as much as you can from the nursery about their growing conditions

The decision of which plants to include in your herb garden is based on what you and your family like to eat, what design style you choose, what the herbs will be used for, and your climate.

Although many herbs adapt to a wide range of environments, you will be heading for disappointment if you choose plants that are wildly unsuitable for your climate. Lavender, which likes dry weather, won't do too well in a humid, damp environment. Assess your climate and choose herbs accordingly. Selecting a theme for your herb garden can also help you decide which plants to grow. Some examples of themes are: Mediterranean, Italian, scented, Mexican, Shakespearean, medieval or medicinal.

Propagating herbs

When I first began growing herbs I saw a photograph of purple sage in a magazine and, although I hunted everywhere, no local nurseries had seedlings. I ordered some seeds and patiently propagated my precious plants. Harvesting and eating from that sage bush (and now its descendant) still gives me great pleasure.

Many herbs, such as thyme and rosemary, grow best by propagating new plants from cuttings, layering or divisions (see page 23). Some, such as coriander, dill and fennel, grow happily from seeds sown directly in the garden, while others, such as lavender, prefer being started off in a seed tray. As a general rule, most perennial seeds should be sown in seed trays first. Not only do they take quite a while to germinate, it is also better to transplant a perennial exactly where you want it, as it will be there for a while. Seed sown directly in the garden is a bit more haphazard.

A good time to plant and sow most herbs is during spring when the moisture and warm weather will give them a chance to get their roots developed. There are some herbs, such as calendula, which are sown in autumn. Hardier herbs also do well if they are planted out during winter. Make sure they are well mulched and they will settle in during the colder months, ready to burst with new growth as soon as it warms up. When herbs are young they need regular water but once established they don't need as much.

Seeds

If you are a novice gardener, plant your herb garden using purchased plants and seedlings. Once you become more confident, try growing some herbs from seed. Most herb seeds take longer to germinate than vegetable seeds. Perennial herbs require even more patience and TLC than annuals as they can take up to a month to germinate. Be aware that the seeds of most herbs have a shorter life span than those of vegetables. Generally, herb seeds are viable for six months to a year, so make sure you are starting with fresh seed and sow more than you need to ensure a good germination rate.

A good time to plant and sow most herbs (such as anise hyssop, opposite) is during spring when the moisture and warm weather will give them a chance to get their roots developed

21 GROWING HERBS

Sowing seeds in seed trays

A variety of containers can be used for starting seedlings. You can either buy seedling pots and trays or recycle something from home. Polystyrene cups and egg boxes are good, as are the inner rolls from toilet paper – the latter two can be planted container and all into the ground, which helps reduce transplanting shock. Use either purchased or homemade seedling soil. I use one part well-sieved compost, one part river sand and one part vermiculite. I then add organic fertiliser, such as Talborne's Seedling fertiliser. Soak the vermiculite first before mixing the other ingredients in and the medium will be wet enough for sowing seeds. If using purchased seedling mix, soak it well before sowing seeds.

Sow seeds according to the depth specified on their packet. Some herbs need light to germinate, others don't, some need to be covered and others left on the surface. There are also herb seeds that need special treatment, such as soaking or chilling overnight. Refer to the herbs' growing information in the A–Z of herbs starting on page 40 for more detail and always read the instructions on the seed packet before sowing. And don't forget to label the seed trays or containers with both the name of the herb and the date.

Place them in a warm, well-ventilated area where you will see them so you remember to keep them moist. In colder, drier weather, cover the seedling trays with clear plastic sheeting. Use sticks or plastic bottles to make a tent above the trays so the plastic doesn't touch the surface. This creates a moist, hothouse environment, which increases germination. When the seeds germinate, turn their containers every few days so the plants don't bend in one direction towards the light.

To keep the seed trays moist, use a spray bottle daily to gently wet the surface, without dislodging the seeds. You can also place the seedling tray inside a larger baking tray or similar container. Fill the baking tray with about a centimetre of water and the moisture will seep up through the bottom of the seedling trays, keeping the growing medium moist. Let the water in the bottom tray dry out completely before refilling it, otherwise the growing medium will stay too wet. This method encourages strong roots to develop as they follow the water.

Once they have grown their first two proper leaves (in addition to the first leaves that came out when they germinated), they are ready to be transplanted.

Sowing seeds directly into the garden

Prepare the area to be seeded so it is smooth and free of any weeds, stones or clumps. Cover the surface with a mixture of finely sieved compost, mixed with vermicompost (earthworm castings) and some Talborne's Seedling fertiliser. Wet it until it's moist but not waterlogged.

Sow seeds one to three times deeper than the size of the seed. Depending on the type of seed and the desired planting pattern, either scatter them evenly across the surface and cover them with sieved compost or drop them into prepared furrows and then close the furrow. Once sown, press the area down firmly using a block of wood. Water the area gently with a mister, so you don't dislodge the seeds. Keep them consistently moist until they germinate and then reduce the watering slightly until the seedlings are established.

Cuttings, layering and division

I once used lemon verbena twigs to hold a drip irrigation hose in place between my tomatoes. A few months into summer, I spotted a very healthy looking lemon verbena plant pushing its way up amongst the tomatoes.

Plants want to grow – we just need to give them the right environment and conditions. Propagating herbs from cuttings, layering or division is the most cost-effective way of bringing new plants into your herb garden; the only price is your time and labour.

Many herbs propagate quite easily using these simple, yet rewarding methods. You really do feel like you are connecting with nature when you conjure a new plant out of a stick!

Cuttings

There is something very empowering about pulling a stem off an established plant, potting it up and creating a new plant from it. My earliest memories of this plant magic is of Mom picking slips from friends' gardens, wrapping them in wet tissues and bringing them home to be transformed into new plants for her garden. When I started becoming interested in doing it myself, I knew instinctively what to do as I had watched her so often.

Many herbs, such as oregano, lavender, pelargonium, rosemary, thyme and sage can be propagated from cuttings. This is an ideal way to keep frost-sensitive plants from season to season or to expand your herb collection. Some perennial herbs last only about five years and by taking cuttings from older plants you make sure you keep the identical plant in your herb garden.

Cuttings are taken from new growth and placed in growing medium until roots grow. Take several cuttings as they won't all grow successfully.

- First prepare your growing medium and fill some containers. The growing medium should be porous but also retain moisture. A good mixture is equal quantities of coarse sand and vermiculite. Wet the growing medium well.
- In spring or early summer pull off a section of new growth, about 10–15 cm long, just below a node. (The nodes are where the leaves join the stem. They contain cells that are primed to make new growth and when they are inserted into a growing medium, instead of making new leaves, they make new roots.) Pull it downward off the stem, taking a bit of the stem's outer layer with it, creating a kind of 'heel'.

Propagating herbs from cuttings, layering and division is the most cost-effective way of bringing new plants into your herb garden

- Trim the leaves off the cutting, leaving two or three at the top. If they are large, leave only two on the stem and cut them in half. (This is so the plant does not waste any of its energy on extra leaves and instead sends it to the roots. It also prevents it from losing too much moisture through its leaves. Enough leaf surface needs to remain for photosynthesis to take place to keep it alive.)
- To encourage root growth, dip the end of the cut section in rooting powder. Available at nurseries, this comes in three strengths: No. 3 is for hardwood stems (such as rosemary), No. 2 is for semi-hardwood stems (such as thyme) and No. 1 is for softwood stems (such as basil). Although this is a synthesised product, the actual hormones are the same as they would be in the plant. If you are averse to using a synthesised product, try using honey instead. It contains elements that resemble growth hormones as well as sealing off the cut cells, helping prevent moisture loss. The sugar in the honey also provides food. It works best for semi-hardwood and hardwood cuttings.
- To prevent the rooting powder or honey being brushed off, first use a stick to poke holes in the growing medium, about 3 cm deep. Insert the cut end of the stem into the hole and firm the growing medium around it.
- To help increase the humidity, cover the containers with plastic, making sure it doesn't touch the soil.
- Keep the containers in a warm place with light. Don't place them in direct sunlight otherwise they will dry out. It should take eight to ten weeks for new roots to form and then the new plant can be transplanted into the ground.

Layering works well with herbs that have woody but flexible stems, such as wall germander (above), sage or lavender (opposite)

Layering

Layering is similar to taking cuttings, except you don't cut a stem off the plant. This method encourages a stem or shoot to form new roots while it is still attached to the mother plant. It works well with herbs that have woody but flexible stems, such as rosemary and sage.

In spring, look for a section of new growth, close to the base of the plant. Strip the leaves off the stem, leaving some at the growing tip. Dig a furrow, the length of the stripped stem, about 6–8 cm deep, and place the bare stem in it, leaving the leafy tip of the stem sticking up. You might need to use a metal hoop to pin the stem firmly to the ground. Cover the bare stem with soil and mulch it. Keep it moist and new roots will start to grow. It can take a few months to establish its roots. When it is strong enough, cut it away from the mother plant and transplant the rooted plant to its new home.

Division

This is exactly what it sounds like – dividing a plant up to create new ones. Spring or autumn are both good times to divide plants. Many herbs can be propagated using this method. Choose plants that are becoming too big and bushy for their space. You will

ABOVE: Maintenance of herbs once they are established is relatively hassle free.

need a sharp spade and perhaps a pair of secateurs to cut the roots apart. Either dig the whole plant up and then firmly push the spade from above, down into the roots, cutting off new sections that have both stems and roots. Or leave the plant in the ground and firmly push the spade down into it, cutting off a section with roots and stems. Dig up the new section, leaving the remainder of the plant where it is. Transplant the divided section to its new home.

Maintenance

Maintenance of herbs once they are established is relatively hassle free. They are not as demanding as vegetables and, as many are perennial, they don't remove as many nutrients from the soil. Keep herb beds well mulched with leaves, compost, hay or grass clippings. Mulch retains moisture, prevents water runoff, provides food and homes for beneficial organisms and prevents weeds from growing.

Feeding

The first step in providing nutrients for your plants is to maintain good-quality soil by regularly adding compost and organic matter. Drought-tolerant, perennial herbs, such as lavender, rosemary, thyme and oregano, don't require much feeding other than the addition of a balanced, organic, slow-release fertiliser in spring. Herbs such as basil and parsley benefit from being given a slow-release fertiliser every four to six months. Don't make the mistake of overfeeding herbs. This leads to unhealthy growth stimulation, which will attract harmful insects and encourage disease. It also reduces the concentration of oils that give herbs their flavour, their scent and much of their medicinal value.

Watering

The amount of water required depends on the herbs. Coriander and rocket will need more regular watering than a large, established rosemary or lavender bush. Group herbs together in beds according to their watering needs – keep herbs that need more water in the same bed. It is better to water deeply less often than to give a little water every day. Drip irrigation or soaker hoses laid in the beds are the most water-wise methods.

Trimming

Herbs need regular haircuts. Many herbs tend to become straggly and woody if they are not cut back and kept trimmed. Trimming will increase the life span of a perennial by continually encouraging new growth. With annual herbs such as coriander and basil regular harvesting prevents flowers forming and encourages the plant to produce much larger leaves. As soon as an annual is allowed to flower, all of its energy will go towards making flowers and hence seeds – and the leaves become smaller and more insignificant.

Weeds

If you mulch regularly, you should not have too much of a problem with weeds. When starting a new bed, cover the surface with a thick layer of wet newspapers and then top it with a layer of compost. Any weeds that make it through this can easily be pulled out. Keep beds as full as possible, spacing plants so they all just touch one another when fully grown. This will crowd out any weeds. Use herbs such as chamomile and creeping thyme as ground covers to prevent weeds from growing.

Don't make the mistake of overfeeding herbs. This leads to unhealthy growth stimulation which reduces the concentration of oils that give herbs their flavour, their scent and much of their medicinal value

Herbal companions

One of the reasons I choose to plant many of my herbs in amongst my vegetables is that this makes my life easier as a gardener. If we simply plant certain herbs in our garden, they will do our work for us. Now that is the kind of gardening I love!

There are many plants that do the work of insecticides by repelling pests. Others are magnets for beneficial insects, inviting bees, wasps, spiders, butterflies and many other predators and pollinators into our gardens. And there are endless combinations of plantings, where one plant benefits another. This is what companion planting is all about.

Pest repelling

Many strong-smelling herbs repel harmful insects and by growing these throughout our vegetable gardens they become defence weapons, confusing the enemy and preventing them from settling in for a feast. These herbs include African wormwood, feverfew, lavender, mint, pyrethrum, rosemary, sage, scented pelargonium and tansy. Most of these don't mind being trimmed regularly, which is lucky as I am constantly cutting off leaves to use as insect-repelling mulch and more.

Birds will notice a newly prepared area and target it. Seedlings, used to being in the semi shade, can wilt quickly when transplanted to a new home in full sun. Cutworms lie in wait under the soil, ready to devour the tender transplants, and slugs and snails make a beeline for the young green growth.

Luckily there is a herbal solution to all of these potential dangers. Herbs with strong stems and fragrance, such as African wormwood, bay, lemon verbena, rosemary or sage, can be used to provide multiple defences. Break off a selection of twigs from any of these bushes and strip off the bottom leaves. Scatter the leaves as mulch in amongst the seedlings. This camouflages a newly prepared area from birds and helps deter slugs and snails. Push the base of the twig into the ground next to the stem of the seedling. This protects the vulnerable stem from cutworms. Bend or break the top bushy bit of the twig and position it like an umbrella to block the seedling from hot afternoon sun. By the time the leaves of the protective twig have completely withered, the seedlings should be strong enough to withstand the heat. One plant providing multiple protection – that is the kind of herb I like.

Trap herbs

If you can't deter a pest then plant a sacrificial herb. A trap herb is one that lures pests away from your vegetables. Nasturtiums, for example, attract aphids away from other plants because they are more attractive to the pests, and dill will lure tomato hornworms away from tomatoes.

Many herbs, such as borage (above) and violas (opposite), are magnets for beneficial insects, inviting bees, wasps, spiders and butterflies into our gardens

ABOVE: Fennel flowers attract the all-important pollinators, such as bees and butterflies.

Attracting beneficial insects

The flowers of many herbs attract the all-important pollinators, such as bees and butterflies, as well as the terminators – or bug-eating insects – such as wasps, spiders, ladybirds and assassin bugs. Examples of herbs that attract these insects are angelica, borage, buckwheat, bugle, calendula, chamomile, coriander, clover, dill, echinacea, fennel, garlic, goldenrod, lavender, lemon balm, marjoram, mustard, nasturtium, parsley, rocket, thyme and yarrow. Many of the strong-smelling herbs that repel harmful insects have beautiful flowers to attract the beneficial ones: feverfew, lavender, mint, pyrethrum, rosemary, sage, scented pelargonium and tansy are some examples.

Other beneficial herbs

Numerous herbs have a beneficial effect on the soil and on other plants nearby. Some, such as fenugreek and clover, do this by adding vital nitrogen to the soil. These herbs are members of the legume family, all of which have a symbiotic relationship with bacteria in the soil. The bacteria take nitrogen gas from the air in the soil and feed it to the legumes. In exchange, the plant provides carbohydrates to the bacteria. When these herbs are cut back and the trimmings used as mulch, the stored nitrogen is returned to the soil in a usable form for other plants.

Other herbs add nutrients to the soil by mining them from deep below the topsoil, where most vegetables and herbs can't reach. Comfrey is a good example, with its deep roots sucking up potassium, calcium and other nutrients and accumulating them in its leaves. Comfrey leaves contain two to three times more potassium than kraal manure. Alfalfa and yarrow are two other examples of herbs that give nutrients to their neighbours.

A further benefit of some herbs is what I call their shape-shifting properties. Harmful insects are programmed to recognise the particular shape of their delectable targets. By planting a companion herb close to the vulnerable vegetable, its shape is disguised. A good example is planting a rambling nasturtium, with its saucer-shaped leaves, around cabbages. This hides the easily recognisable shape of the cabbages from the cabbage moth as it flies over looking for their distinctive silhouette.

Then there are the herbs that are simply good bed mates. Examples are: basil with tomatoes, mint with eggplant, horseradish with potatoes, and yarrow, which improves everything near it. I am constantly discovering new uses in my garden for herbs in an ongoing process of trial and observation. If you don't have the space to grow these wonderful protective herbs in amongst your vegetables, plant them in pots and move them to where they are needed.

Some herbs add nutrients to the soil by mining them from deep below the topsoil, where most vegetables and herbs can't reach. Comfrey (above), with its deep root system, is a good example of this

Pests and diseases

Luckily herbs aren't too bothered by many pests or diseases, especially once they are well established. Most problems will occur when a herb is either planted in the wrong conditions or given the incorrect amount of water. Overwatering Mediterranean herbs or planting them in a damp, soggy environment could lead to fungal problems. Insufficient air circulation can also lead to fungal problems, such as rust.

Red spider mite can occur on herbs such as thyme, sage, rosemary, oregano and marjoram, especially in hot, dry conditions. Regular, deep watering will help prevent this as well as encouraging ladybirds that eat spider mites.

Some herbs, such as basil, nasturtiums and chives, can be bothered by aphids. If you have a severe infestation, blast them off with a hose or spray with a mixture of garlic and sunflower oil.

Prevent any of these problems happening by choosing the right herbs for your garden and by planting as much variety as possible to encourage a balance of both the good and the bad guys.

'I have no space' is not an excuse when it comes to growing herbs. By planting herbs, such as lavender (above), in containers you can literally have a herb garden on your kitchen doorstep

Herbs in containers

'I have no space' is not an excuse when it comes to growing herbs. Even if you grow only a few pots of herbs on a sunny windowsill you will be saving yourself money and opening a window onto a whole new way of cooking and using edible plants. With containers you can literally have a herb garden on your kitchen doorstep.

Growing medium

A mistake often made when growing herbs in containers is to use the wrong growing medium. A growing medium's functions are to provide water, nutrients and support for the plant's roots. The medium needs to drain so the roots don't sit in wet soil but it also needs to retain sufficient moisture so they can access nutrients and don't dry out. Using just compost or digging up some garden soil won't result in happy plants. Either buy a good-quality organic potting soil or mix your own. Even when I purchase potting soil, I usually tart it up by adding some of the ingredients below. A good recipe includes a mixture of:

- Disease-free topsoil. To provide nutrients, density and natural organisms.
- River sand. Its large, coarse grains provide air pockets as well as weight to tall containers.
- Sieved, nutrient-rich compost. If you make your own compost using manure and nutrient-rich herbs such as yarrow, the compost will provide nutrients as well as retain moisture. Earthworm compost is an excellent addition.
- If your compost doesn't contain manure, then some well-rotted kraal or horse manure can be added to your potting soil.
- A further moisture-retaining ingredient should also be added. My favourite is vermiculite as it holds air, water and nutrients. Coir or even torn, scrunched up newspapers can be used.
- And finally, add a slow-release, organic fertiliser to the mix.

Containers

Many containers can be used for growing herbs. My friend Allison collects old biscuit tins as planters for her herbs. Your container garden can be classic (a selection of terracotta pots) or quirky (recycled wheelbarrows and cracked old baths), funky (brightly coloured plastic bowls and jugs) or modern (sleek aluminium containers). Whatever style you choose, it should fit your budget and style. Be aware that the smaller the container is, the quicker it will dry out.

One advantage of growing herbs in containers is they can be moved around as the seasons change to take advantage of the sun. If the container is large and you do plan to move it, put wheels underneath it before filling it.

Container plants need to be fed more regularly as the nutrients are washed out of the growing medium

Remember that plants benefit from good air circulation, so don't cram too many containers together. Create varying heights by using different size containers or by raising them up on bricks or blocks of wood.

Maintenance

As with herbs growing in the ground, herbs in pots need food, water and regular trimming. Container plants need to be fed more regularly as the nutrients are washed out of the growing medium. Use a dry organic fertiliser every three to four months. A liquid organic fertiliser should be used every month on annual herbs in summer and once every three months on perennials.

Containers dry out more quickly and need to be watered more often than plants in the ground. Get into the habit of checking often. Installing an irrigation system will ensure you don't kill your plants. A simple drip pipe wound through the pots will work. If your irrigation system has spray nozzles, they often waste water by spraying over the edges of the pot. I saw a very simple yet clever solution to this in the container garden of my friend Sam. He positions a cut-off plastic bottle over the top of the nozzle, which directs the spray downwards into the pot.

Keep container herbs in good shape by regularly trimming them. This not only keeps them healthy, it also maintains the overall balance in your containers.

HARVESTING

The efficacy of herbs can be affected if they are not properly harvested. If you are picking a few herbs to make a quick herbal tea or to add to a dish, it doesn't really matter when you pick them, but if you are going to dry them, it is best to pick on a dry day, after the morning dew has evaporated. Harvest the herbs at their peak of maturity, when the concentration of active ingredients is at its highest.

- Deciduous or annual leaves should be picked just before flowering.
- Evergreen perennial leaves can be picked year round.
- Cut whole flowers when they are fully open.
- If using all the parts of the plant, harvest while it is flowering, picking a mixture of leaves, flowers and seeds.
- Most roots are harvested in autumn, once the plant has died down. (An exception is dandelion – its roots are best harvested in spring.)
- Harvest whole seed heads with some stalk, when they are almost ripe.

Herbs such as mint (opposite) dry well, retaining their flavour and efficacy

Preserving

Drying

Many herbs dry well, retaining their flavour and efficacy. When using herbs you have grown and dried yourself, you can be assured they are 100% organic without any pesticides.

- Herbs with a lower water content, such as oregano, rosemary and bay, are well suited to drying.
- Dry flowers, seeds and leaves quickly, out of direct sunlight, in a warm, well-aired spot. (The kitchen might seem an obvious spot but is often full of condensation and cooking smells.)
- Shake whole flowers to remove any obvious dirt or insects and dry them whole, spread out evenly on a piece of paper or a wire rack.
- For large-leafed herbs, such as comfrey, pick and dry the leaves individually.
- For small-leafed herbs, leave them on the stem, gather the stems together and tie them in a bunch.
- Tie seed head stems together and pop them in a paper bag with the stems sticking out. Tie the bag closed around the stems and hang it upside down. Once they are fully ripe, shake the seed heads to dislodge them into the bag.
- Wash roots well and chop them while still fresh. Spread them onto paper and dry them in an oven at 100°C for 2 to 3 hours. Leave them in a warm, dry place until completely dry. Flowers and leaves can also be dried using this method if you want to speed up the process.
- Once they are dry (and make sure they are completely dry otherwise they will go mouldy) store the herbs in an airtight sterilised glass or ceramic container. Either use a dark glass bottle or keep them in a dark cupboard. Most herbs will keep for up to 18 months.

Preserving in salt or sugar

This method of storing herbs and keeping their fresh flavour uses salt and sugar as natural preservatives. Herbs best suited for salt are strongly flavoured culinary herbs such as marjoram, oregano, rosemary, winter savory and thyme. Herbal sugar can be made using fragrant herbs such as lemon verbena, rose-scented pelargonium, anise hyssop, mint and lavender.

To make herbal salts or sugars, pick a colander of herbs, and wash and dry them well using a salad spinner. Strip off the leaves, discarding any woody stems. Using a food processor, chop the herbs until fine. Measure the herbs and for every cup of herbs add two cups of sea salt or sugar. Add half a teaspoon of lime or lemon zest to the salt. Mix together well. Using a blender, finely blend one cup of the mixture at a time until it is smooth and pale green. Store in airtight bottles.

Herbs can be preserved in oil (above) or spread out evenly to dry.

Preserving in honey
Making herbal infused honey is a versatile way of preserving and using herbs, often with great medicinal benefits. See page 190 for a description of how to make and use herbal honey.

Preserving in oil
Many herbs store well in oil, with the additional advantage of creating a herb-infused oil. Simply tear up leaves or flowers, layer them in a sterilised bottle and cover with olive oil. Poke with a sterilised spoon to remove any air bubbles. As long as the herbs remain covered with oil, they will last for months. For more on infused oils see page 192.

Freezing
Freezing works well with herbs that have a high water content, such as basil and coriander. To freeze whole leaves, place them in a single layer on paper towel laid on a baking sheet. Pop this into the freezer and when they are completely frozen, store them a zip lock bag. Or blend the herbs with a little olive oil until you form a thick paste. Freeze this in ice trays until solid, remove the blocks and store in zip lock bags. Add whole cubes to soups, or defrost them to use in salad dressings and tea, or to mix into a lotion.

A–Z OF HERBS

Herbs are versatile plants, adding diversity, texture and scent to our gardens. They are also rewarding, providing bountiful harvests for a multitude of purposes. On top of all this, herbs are some of the easiest and most tolerant plants to grow.

In the following A–Z of Herbs, you will find information on how to grow, harvest and use some of my favourite herbs. Starting on page 176, the Herbal Pharmacy gives detailed explanations of how to make the various preparations that I refer to in this section, from simple infusions and decoctions to more complicated tinctures, body creams and balms. The Herbal Pharmacy also includes information on the equipment and ingredients you will need to make your own herbal preparations. The Herbal Recipes chapter, starting on page 198, has over a hundred fun and useful recipes for making the most of the herbs in your garden, whether for cooking, healing, cleaning, pet care or feeling good.

A

AFRICAN POTATO
Hypoxis hemerocallidea

Indigenous to South Africa, the African potato has been used for centuries by traditional healers. In the last few decades it has gained prominence for its ability to strengthen and perk up the immune system. It is only during the last 20 years or so that it has been called African potato, which is a bit of a misnomer as it grows from a corm (a fleshy food-storing underground stem) and not from a tuber like the regular potato.

Growing
- Hardy perennial
- Well-drained, fertile soil
- Full sun
- 40 cm high by 30 cm wide

Harvesting
Harvest corms in autumn to early winter from established second-year plants. Dry well before slicing or grinding to a powder. Store in sealed, dark bottles.

Parts used
Corm

Uses
- Builds immune system
- Tonic
- Antiviral
- Detoxifier
- Reduces blood pressure

Cautions
This herb should be taken under the guidance of a trained herbalist.

In the garden
With its bright yellow, star-shaped flowers, this is an attractive and hardy plant that attracts beneficial insects to the garden. It is drought tolerant and likes full sun in well-drained soil. Buy young plants from a specialised herb nursery. Enrich the soil with compost before planting. Mulch with compost a couple of times a year, especially in winter when it goes dormant. It is happier in areas with summer rainfall.

Growing in containers
The African potato can be grown in a large container with well-drained growing medium. Make sure it is kept well watered throughout summer.

Healing properties
This plant has the ability to increase the functioning of white blood cells, which control and regulate the immune system, and to purify the blood. It has been used to effectively treat a range of illnesses from colds and flu to tuberculosis. It has even been claimed by some to be an effective treatment for HIV/Aids. It is effectual against asthma, allergies, high blood pressure, cholesterol, diabetes and arthritis.

How to use
Infusion: To build and strengthen the immune system; to treat diabetes, arthritis, urinary ailments, allergies and asthma; to reduce blood pressure and cholesterol; to speed up the healing of convalescing patients.
Decoction: As for infusion, but stronger.

Other uses
The leaves and corm can be used as a black dye.

A

AFRICAN WORMWOOD
Artemisia afra

This tall grey-green shrub is a valuable addition to the herb garden. Indigenous to South Africa, it is a member of a large family of plants containing more than 400 species. This bitter, strong herb has been used for centuries for its medicinal properties. I find it one of the most useful plants in my vegetable garden because of its pest-repelling qualities.

Growing
- Hardy perennial
- Light, well-drained soil
- Full sun
- 2 m high by 2 m wide

Harvesting
Harvest young shoots and leaves from midspring through to the end of summer. Leaves can be dried for later use.

Parts used
Leaves

Uses
- Drying and cooling
- Digestive tonic
- Uterine stimulant
- Expels worms
- Antibiotic
- Antiseptic

Cautions
Do not take when pregnant or breast-feeding. Do not take habitually or in excess.

In the garden

African wormwood grows easily from seedlings or cuttings and is unfussy about soil. It likes full sun and once established it becomes a large, bushy and drought-resistant plant. It needs little more than an occasional cutting back and can grow up to 2 m tall. In early spring, cut it back by two thirds to encourage new growth.

It has very strong-smelling leaves, which few bugs can withstand. African wormwood performs a multitude of functions in the garden and is my go-to plant when I have vulnerable seedlings or seeds needing protection. Have a look at page 29 for some of its protective uses. You can scatter torn-up African wormwood between the leaves of cabbages, cauliflower or broccoli to repel leaf-eating worms. Place leaves underneath strawberries to protect them from snails. It is also a good shrub to plant in amongst fruit trees as it helps repel harmful insects.

Growing in containers

Grow African wormwood in a large container and keep it trimmed.

Healing properties

As the plant's common name suggests, a popular use of the herb is to expel worms. However it has a wide range of medicinal uses from treating digestive ailments to relieving bronchial problems, fevers and colic.

How to use

Infusion: For slow digestion, colic, poor appetite and other gastric problems, to treat bronchitis and colds and to expel worms.
Tincture: As for infusion, but stronger.
Compress: To soothe bruises and insect bites.

Other uses

African wormwood is very useful as an insect repellent in the house and for pets. Sprinkle dried African wormwood around ant holes and mix with corn flour to sprinkle on dogs to chase away fleas (see page 254 for a recipe). Place a dried leaf in storage containers of dried beans and flour and other grains to keep weevils out.

A

ALFALFA
Medicago sativa

Grown worldwide as fodder for livestock, alfalfa is one of our oldest-known cultivated crops. With its edible mauve flowers, it is a surprisingly attractive plant to add to a herb garden. In the realm of magic, alfalfa has a reputation of being a lucky herb, bringing money into the home and warding off poverty.

Growing
- Hardy perennial
- Well-drained soil
- Full sun
- 1 m high by 60 cm wide
- Deep root system

Harvesting
Harvest the flowering tops and young leaves from spring through to late summer. These can be dried for later use.

Parts used
Leaves, flowers, young shoots and sprouts

Uses
- High in chlorophyll and nutrients
- Alkaliser
- Detoxifier
- Lowers cholesterol
- Relieves arthritis
- Encourages hair growth
- Regulates blood sugar

Cautions
Avoid daily consumption of alfalfa during pregnancy.

In the garden
This is one of my favourite herbs: simple to grow, pretty to look at, healthy to eat and beneficial to other plants. You need only one or two plants to enrich a medium-sized herb garden. A perennial, it is easily grown from seeds or seedlings in full sun. Drought resistant and hardy, it dies back in winter in cold areas but will spring up when the weather warms, quickly reaching a metre high. Alfalfa benefits from being cut quite often and this encourages roots to slough off, adding valuable organic matter to the soil. The trimmings can be used as nutritional mulch or added to the compost. The purple flowers attract beneficial insects, particularly bees.

Growing in containers
Alfalfa does not grow well in containers as it has a very deep root system.

Healing properties
Recognised as a medicinal plant for thousands of years, alfalfa provides a multitude of healing benefits. It is an extremely nutritious plant, containing a wider variety of nutrients, minerals, trace elements and vitamins than many other plants (it is particularly high in vitamins A, B, C and K). Alfalfa is rich in silica, which helps hair that splits easily or is thinning.

How to use
Infusion: (leaves and flowers) To lower cholesterol, relieve arthritis, regulate blood sugar, reduce acidity, improve digestion and detoxify the blood. To treat urinary, kidney and prostate problems; for thinning or splitting hair; irregular menstruation and menopause symptoms.
Tincture: (leaves and flowers) As for infusion, but stronger.
Juice: Juice young leaves, flowers and sprouts. Drink as a tonic when required.

Culinary uses
The flowers, leaves and young shoots are a vitalising tonic in salads. Sprout seeds for a nutritious addition to salads and sandwiches. Juice leaves to add to smoothies and drinks.

Other uses
With a wide-spreading and deep root system, alfalfa is a nutrient-accumulating and soil-enriching plant. Part of the legume family, it also fixes nitrogen.

A

ALOE VERA
Aloe vera

From ancient China, Greece and Rome to modern-day India, aloe vera has been used through the centuries as an extremely effective healing plant. The beauty of this plant is that you can just break off a piece and rub the gel right onto your skin – ideal for a hard-working gardener's hands! In folklore tradition, it protects the home against accidents and wards off evil influences.

Growing
- Semi-hardy, evergreen perennial
- Light, well-drained soil
- Full sun to partial shade
- 1 m high by 60 cm wide

Harvesting
Harvest larger older leaves from the lowest part of a plant older than two years.

Parts used
Leaves and gel inside leaves

Uses
- Soothing, cooling, astringent, antiseptic and antifungal, all-purpose wound healer
- Blood cleanser
- Relieves inflammation

Cautions
The outer yellow sap is a powerful laxative and can cause diarrhoea and cramps if too much is taken internally. It should not be taken internally when pregnant.

In the garden
Easy to grow, with flowers that provide nectar for bees in winter, this is a valuable addition to any garden. Aloe vera likes full sun to partial shade and well-drained soil. It can be propagated by transplanting the shoots that grow from its base once it is established. Allow the shoot to dry for a day before planting in well-drained soil. Drought resistant and semi-hardy, it does not do well in very cold climates. Plants older than two years have stronger healing properties.

Growing in containers
Aloe vera can easily be grown in containers – even as a houseplant on the windowsill. The growing medium must have good drainage. Do not over-water.

Healing properties
The leaf is made up of several layers: the outer rind, a thin inner layer of yellow sap and a large inner section of gel and mucilage. This inner thick, soothing and cooling section of the leaf heals burns, stings, rashes, wounds and sunburn. It is an excellent moisturiser, especially for sensitive skin, and reduces scarring. It also effectively treats fungal infections. Taken internally, the bitter gel stimulates bile, aiding digestion and stimulating the appetite and the liver. The yellow sap from the leaves can be taken internally to treat constipation.

How to use
Fresh: Split the leaf and apply the gel directly to the affected area. The cut leaf will seal over and can be used again. To treat burns, sunburn, insect bites, inflamed and itchy skin (especially good if you are itchy from handling plants), rashes, shingles and fungal infections of the skin. Apply gel to the outside of the eyelid to treat conjunctivitis. Add fresh gel to moisturising creams for dry skin.
Infusion: (gel) To treat slow digestion, anaemia, stomach ulcers, constipation and liver congestion, and as a tonic.
Tincture: (whole leaf) To treat stubborn constipation and liver congestion.
Steam inhalation: Add gel to boiling water and inhale the steam to ease bronchial congestion.
Ointment: Split a leaf open and scrape the gel into a pan. Simmer slowly until it thickens. Store in a sterilised bottle in the fridge for a month or in the freezer for up to three months. To treat aches, pains and swollen joints.

Ancient Arabs used their bare feet to crush aloe vera leaves and extract the gel

A

ANGELICA
Angelica archangelica

This interesting plant has been associated with witchcraft and sorcery for centuries. Even its name indicates that it will call on the angels to protect from evil. It is an old-fashioned English cottage garden plant. The taste and smell of angelica transports me back to childhood Christmas parties, where candied angelica was served as sweets.

Growing
- Hardy biennial
- Well-drained, moist soil
- Dappled shade and cool climate
- 1.5 m high by 1 m wide

Harvesting
Cut leaves and stalks once they are big enough to use. Dig the roots out at the end of the second year. The leaves, roots and stems can be dried for later use.

Parts used
Leaves, flower buds, stems and roots

Uses
- Warming and stimulating
- Relieves inflammation
- Relieves spasms
- Pain reliever
- Detoxifier
- Strengthens immune system
- Expels phlegm

Cautions
Not to be taken internally by pregnant women or diabetes sufferers.

In the garden
Angelica is a tall biennial plant with long hollow stems and showy greenish-white flowers in its second summer. As it can grow well over a metre tall, it should be positioned to the back of the herb garden. Angelica grows easily from seeds and seedlings. The seeds are viable for three to four months so this is an option only if you harvest your own. Transplant or sow the seeds directly in a moist area with dappled or afternoon shade. It grows well on the southern side of an elderflower or similar tree. It won't do well in a hot and dry climate. It benefits from ground cover or a well-mulched surface to keep its growing environment moist. The flowers attract beneficial insects. Cut the seed heads before they start falling and dry them prior to storage. Plants die once the seed has matured but this can be delayed by removing the young flower stems.

Growing in containers
Angelica can be grown in a large container but needs staking or support.

Healing properties
All parts encourage perspiration, stimulate appetite and are used to treat poor circulation, digestive problems and bronchial ailments, particularly as an expectorant. Angelica is a diuretic and detoxifier, good for treating cellulite and fluid retention. The roots have anti-inflammatory, antispasmodic and pain relieving properties. The roots are also used to treat various gynaecological ailments, from menopause symptoms to menstrual cramps. It is used externally to treat painful joints.

How to use
Infusion: (leaves) To treat colic and flatulence. To increase circulation, detoxify system, reduce cellulite and fluid retention, and strengthen immune system. To reduce indigestion and improve appetite.
Decoction: (roots) For menopause symptoms and menstrual cramps.
Tincture: (leaves) To treat digestive upsets, asthma and other bronchial problems, particularly as an expectorant for stubborn or infected phlegm.
Tincture: (roots) For bronchial and digestive problems and as a liver stimulant.
Cream: (leaves) Apply to minor skin irritations and to reduce cellulite.
Compress: (roots) Use fresh roots or soak a cloth in diluted hot tincture or decoction and apply to painful joints or the temple for migraines.

Culinary uses
Add angelica flower buds and blanched young shoots to salads. Thicker stems and leaves can be added to tart dishes such as rhubarb and gooseberries to sweeten them. Add leaves and stems to poaching liquid for seafood or to marinades. Layer the stalks in a roasting pan under fish before grilling. Wrap fish in angelica leaves and bake. The stalks can be preserved with sugar and eaten as a refreshing and vitalising snack.

A

ANISE HYSSOP
Agastache foeniculum

Part of the mint family, this delicious-smelling plant is a sweet-scented addition to a fragrant herb garden. This is a cleansing herb and it was used by some Native American tribes as incense to purify sacred spaces or to clear a guilty conscience.

Growing
- Hardy perennial
- Moist, fertile soil
- Full sun to semi shade
- 60 cm high by 40 cm wide

Harvesting
Harvest leaves throughout summer and flowers when they are in full bloom. It dries well for later use.

Parts used
Leaves and flowers

Uses
- Cooling
- Antiseptic
- Encourages sweating
- Uplifting

In the garden
As with any mint, anise hyssop is easy to grow. It can be propagated from an existing plant by taking cuttings in spring or dividing an older plant (see page 23). It likes moist soil and full sun to semi shade. It will die back during winter in areas with frost but springs back up as soon as the weather warms up. Bees go crazy for the beautiful spikes of lilac flowers that come out in summer.

Growing in containers
Anise hyssop grows well in a medium to large container. Keep it well watered during drier periods.

Healing properties
The Native Americans used this as an uplifting tea to alleviate anxiety and depression. It was also used to treat coughs and colds and to induce sweating. It is a useful herb to add to other teas as it does not have strong medicinal properties, however it adds a delicious aniseed flavour, particularly good for bitter teas. It can be added to any cough syrup to enhance the flavour and increase its efficacy.

How to use
Infusion: (leaves and flowers) To lift depression and reduce anxiety. To encourage sweating and treat coughs and colds.
Syrup: (leaves and flowers) To treat coughs.

Culinary uses
The leaves and flowers add an anise flavour to salads, fruit, fish and vegetable dishes. Infuse sugar syrup with the leaves as a poaching liquid for fruit or as a base for ice cream. Freeze the flowers in ice cubes and add to cocktails.

B

BASIL
Ocimum basilicum

Growing
- Tender annual
- Well-drained, fertile soil
- Full sun, doesn't mind afternoon shade
- 45 cm high by 30 cm wide

Harvesting
Harvest leaves once the plants are at least 20 cm high. When the flower buds form, pinch them off at least two nodes down to encourage leafier growth. It does not dry well and is better preserved in oil.

Parts used
Leaves and flowers

Uses
- Restorative, refreshing and warming
- Antioxidant
- Antiseptic
- Expels phlegm
- Prevents vomiting
- Antidepressant

No herb garden should be without basil. It is one of the most popular and widely grown herbs. This is a herb long associated with love, and a man will supposedly fall in love with a woman if he accepts her gift of basil.

In the garden
There are more than 60 species of basil, ranging from tender annuals to hardier perennials with a strong, camphor-like scent. The main culinary basil is sweet basil, which comes in various shapes (from small purple-leafed bushes to large green-leafed mammoth varieties) and flavours (from citrus-scented to liquorice-flavoured ones).

Basil likes a protected site with well-drained soil. It prefers full sun but doesn't mind some afternoon shade in summer. It can be grown from seeds or seedlings. Sow basil seeds in modules and transplant the seedlings into larger pots, so they can grow bigger while waiting for the weather to warm up. Basil likes hot weather – I usually wait until early October before planting the seedlings out. Basil and tomatoes are good companions in the garden as basil improves the flavour of tomatoes. Basil also repels insects and flies and helps to prevent mildew on cucurbits (this plant family includes pumpkins, squashes, marrows, cucumbers and melons). Don't plant rue and basil near each other – they will both suffer.

Growing in containers
Basil grows easily in containers and can be grown in pots on a windowsill.

Healing properties
Basil is an uplifting herb, good for treating depression and fatigue or to improve concentration. It also eases colds, asthma, coughs and sinus congestion. As an antispasmodic it prevents vomiting and is a good tonic and soother for the digestive system. It strengthens the immune system, reduces inflammation and kills parasites.

How to use
Fresh: Rub leaves on insect bites or rashes to reduce itching and swelling.
Infusion: Mix with honey and lemon for an uplifting drink, or infuse on its own to treat digestive problems and reduce nausea.
Tincture: For coughs and colds, to prevent nausea and to relieve stress and depression.
Juice: Mix with an infusion of cinnamon, cloves and ginger to reduce fever and nausea.
Syrup: Mix with honey to relieve a congested chest or tight cough.
Steam inhalation: For relief of headaches, depression, fatigue or nervous tension, combine with lavender and peppermint essential oil for better effect. Place fresh leaves in a steamer for relief of sinus congestion, a head cold or nausea.
Essential or infused oil: Add to a bath to uplift nervous exhaustion, depression or mental fatigue (mix with a carrier oil first).

Culinary uses
Basil has a strong and unique flavour. It is best to add it right at the end of a dish to make the most of its flavour. It is most commonly used in Italian dishes, paired with tomato. However its robust flavour makes it the ideal herb to accompany a wide range of dishes.

Cautions
Use this powerful herb sparingly as its stimulating qualities could have the opposite effect if used in excess. Basil oil may irritate sensitive skin. Avoid if you have liver problems. During pregnancy use basil with care and do not use basil essential oil.

B

BAY TREE
Laurus nobilis

Growing
- Hardy, evergreen perennial
- Light, well-drained soil
- Full sun
- 3–12 m high by 6 m wide

Harvesting
Harvest leaves when needed. Bay leaves dry well.

Parts used
Leaves

Uses
- Protective
- Warming
- Insect repellent
- Relieves spasms
- Pain reliever
- Digestive

Cautions
It can cause skin irritations. Do not use bay leaf oil during pregnancy.

This stately tree, native to the Mediterranean, has symbolised glory and honour for centuries. The bay tree was sacred to the sun god Apollo and groves of revered bays were planted near Greek temples. During the height of the Roman Empire, branches were gathered and woven into wreaths to honour the great heroes of the time. (The Latin name for bay is *laurus* – hence laurel wreath.)

In the garden
The bay tree in my garden is about 8 m tall. Unfortunately it is on the northern side of my vegetable garden and blocks precious winter sun, so every year its new growth is trimmed. This harvest is a useful addition to my garden, with the long and flexible branches being used to make fences and small tripods and the leaves added to the garden as an insect-repelling mulch.

Grow from seedlings or take cuttings from an established tree (see page 23). As they are evergreen, bay trees are easy to grow, whether they are allowed to reach their full height or trimmed to suit a smaller garden. They like full sun and are quite drought tolerant. If a tree requires trimming, do it in early spring. Young trees will need protection from frost until they are well established.

Growing in containers
The bay tree's upright and dense growth makes it an ideal herb for a pot. It can be shaped into a topiary or allowed to grow in its natural shape. Bays tend to become top heavy so use a large pot with a broad base and shelter the tree from the wind.

Healing properties
Bay leaves relieve migraines. Their antispasmodic properties ease muscular pain and stomach aches, and aid digestion.

How to use
Infusion: To stimulate appetite, relieve colic and aid digestion.
Decoction: Soak aching limbs in a warm decoction.
Steam inhalation: Mix with boiling water and inhale the steam to ease migraines.
Compress: Soak a pad in a decoction and use to relieve aching limbs and joints.
Infused oil: Rub on temples to ease headaches and migraines; add to bath to ease aching limbs and joints.

Culinary uses
Bay is an indispensable ingredient in *bouquet garni*, a bundle of classic herbs used to flavour French dishes. Fresh leaves added to stews and sauces are much stronger than dried. Dried leaves will keep in a sealed container for six months to a year. Use bay branches as skewers to add a delicious flavour to meat.

Other uses
Add dried bay leaves to bottles of rice, beans and other dried goods to deter weevils. In warmer climates, bay trees produce dark purple berries. These are poisonous to eat, however their juice is an effective treatment for venomous insect stings. The leaves take a long time to break down and can be used as insect-repelling mulch. The bay is believed to increase psychic powers and is added to clairvoyant brews. To increase prophetic dreams and visions, either burn the leaves as incense or place them under your pillow.

B

BERGAMOT
Monarda species

Growing
- Hardy perennial
- Well-drained soil
- Full sun to light shade
- 60–80 cm high by 40 cm wide

Harvesting
Pick leaves fresh when needed. For drying, harvest leaves before the plant starts flowering.

Parts used
Leaves and flowers

Uses
- Antiseptic
- Relieves nausea and indigestion
- Antidepressant

Bergamot's name is due to its scent being similar to the bergamot orange, a citrus tree used to make Earl Grey tea. Bergamot has been used for centuries by the Native Americans to treat a variety of ailments. Also known as bee balm, this herb attracts bees and other beneficial insects into the vegetable garden. Lemon balm is also called bee balm, but the two plants are unrelated.

In the garden
Bergamot is easier to grow from seedlings and likes full sun but will tolerate light shade. It is a member of the mint family and, under the right conditions, can be invasive. If it becomes too big, divide the clump in autumn or spring (see page 25). After it has flowered, cut the dead flowers off to encourage further flowering. It has strongly scented leaves that do a good job of repelling harmful insects. It is an excellent companion for tomatoes and all members of the brassica family.

Growing in containers
Bergamot grows well in a large container. Keep the soil moist.

Healing properties
It contains high levels of thymol and its strong antiseptic action is used to treat mouth and throat infections, skin infections and minor wounds. It is also used to relieve nausea, vomiting and flatulence and to lift depression.

How to use
Infusion: (leaves) To ease a sore throat; to treat mouth, throat and respiratory infections; to relieve indigestion, nausea and vomiting; to lift depression. As a wash to treat skin infections and minor wounds.
Poultice: (fresh leaves) To treat skin infections and minor wounds.

Culinary uses
The leaves of bergamot are strongly flavoured, so use sparingly. Use the leaves to flavour jellies, jam and fruit drinks. Add leaves and flower petals to salads. The leaves add a complementary flavour to pork dishes, and bergamot jelly goes well with pork.

B

BITTER ALOE
Aloe ferox

Indigenous to South Africa, this hardy aloe provides much-needed nectar for birds and bees during the winter. Its Latin name *ferox*, meaning 'warlike' or 'fierce', is taken from the sharp spines on the edges of its leaves.

Growing
- Hardy perennial
- Well-drained soil
- Full sun
- Up to 3 m high by 60 cm – 1.5 m wide

Harvesting
Leaves should be harvested when plants reach four or five years old. Use older leaves from lower down on the plant and cut them off close to the stem. The leaf contains an outer layer of bitter yellow sap and an inner fleshy gel. The sap will drain out slowly. Once it has finished draining, cut the leaf open to scrape out the gel.

Parts used
Leaf sap, leaf gel, dried leaves and roots

Uses
- Bitter
- Soothing
- Cooling
- Astringent
- Antiseptic
- Antifungal
- All-purpose wound healer
- Detoxifier
- Laxative
- Relieves inflammation

In the garden
Like all members of this family, bitter aloe is easy to grow. It is a drought-hardy and tough plant that likes full sun. Plant cuttings from an existing plant in well-drained soil.

Growing in containers
Bitter aloe is suitable for a large container with well-drained growing medium. Do not over-water.

Healing properties
Bitter aloe has been used for hundreds of years as a healing plant in South Africa. The bitter yellow juice found just below the surface of the leaf is dried and taken as a laxative. The gel is used to treat digestive problems, heal minor wounds and soothe dry skin. The whole leaf is used to treat eczema, arthritis and eye infections.

How to use
Fresh: Split the leaf and apply the gel directly to the affected area. The cut leaf will seal over and can be used again. To treat burns, eczema, sunburn, insect bites, inflamed and itchy skin, rashes, shingles and skin fungal infections. Add fresh gel to moisturising creams for dry skin.
Infusion: (gel) To treat slow digestion and to ease indigestion. As a detoxifier and cleanser, particularly for the intestines and bowel.
Infusion: (whole leaf) As a wash for infected eyes and to treat eczema.
Tincture: (whole leaf) To treat stubborn constipation.
Ointment: (whole leaf) Cut up a leaf and place in a pan, just covered with water. Simmer slowly until the mixture thickens. Strain and store in a sterilised bottle in the fridge for a month or in the freezer for up to three months. To treat skin conditions, aches, pains and swollen joints.

Cautions
Do not take the sap when pregnant, as it is a uterine stimulant. Do not take sap when breast-feeding as it can cause colic in babies.

B

BLACKBERRY AND RASPBERRY
Rubus fruticosus and *Rubus idaeus*

Growing
- Perennial crown and biennial canes, which bear fruit in their second year; some varieties fruit in the first year
- Well-drained soil
- Full sun to semi shade
- Spreading height and width

Harvesting
Fruit is ready as soon as it can easily be pulled off the plant. Harvest fresh leaves as required and harvest leaves for drying before the plant starts flowering.

Parts used
Leaves, fruit and stems

Uses
- Drying, astringent, cooling
- Uterine tonic
- Digestive
- General tonic
- Diuretic

Cautions
Do not drink during early months of pregnancy as it can stimulate the uterus.

These rambling fruit-bearing bushes can easily be grown in a small garden, provided they are given the right support and kept under control. It is said that if you bathe in raspberry juice, your mate will not stray. This reputation makes sense when you consider how sweet the juice is – yet how tenacious the thorns are.

In the garden
Blackberries and raspberries prefer full sun. However, they will still produce fruit in semi-shaded areas. They like well-drained soil and don't do well with cold wet feet. They will grow in quite poor soil but will produce more fruit in fertile soil. On some varieties, the first year of growth will produce berries. On others the first year grows canes up to 5 m long that produce only leaves. In the second year, fruit-bearing side branches develop. After fruiting, these canes die back. The fruit will begin to form in early summer.

The *Rubus* species of plants can spread vigorously as the canes root themselves whenever they touch the ground. To control their rampant growth, place upright poles around a narrow bed and string wire around them to create a 'cage'. Whenever a cane starts growing out of the fenced area, wind it back inside. Once the designated growing area is full, remove any new growth that emerges outside this area. Keep the new growth thinned out to about 15 canes per square metre. In late winter or early spring, cut the canes back to encourage new growth in spring. Be careful when working with these plants as the canes are covered in little prickles that hook you unmercifully.

Growing in containers
Rubus species are not suitable for growing in containers.

Healing properties
Blackberry leaves are high in tannins, making them an effective remedy for sore throats and upset stomachs.

Raspberry leaves and fruit are rich in vitamins and minerals. They are also full of antioxidants. This is an astringent herb, used to treat digestive disorders, sore throats and mouth ulcers. It is also known as the 'woman's herb' for its many uses for menstruation and during pregnancy.

How to use blackberry
Infusion: (leaves) As a mouthwash for mouth ulcers and gum inflammation; as a gargle for sore throats. Drink to relieve diarrhoea and cystitis.
Poultice: (leaves) To treat wounds and insect bites.

How to use raspberry
Infusion: (leaves): Drink during the last six to eight weeks of pregnancy to strengthen and tone uterine tissues; drink to increase breast milk, as a general tonic, to treat mild diarrhoea and digestive disorders. Use as a gargle for sore throats and mouth ulcers. Use as a wash to clean and treat minor wounds and to soothe tired eyes.
Tincture: (leaves) More astringent than the infusion, the tincture is used to treat minor wounds, skin conditions and inflammation. As a mouthwash to treat infections.
Vinegar: (berries) Add to cough mixtures or use as a gargle for sore throats.

Culinary uses
Both raspberries and blackberries can be eaten raw or made into jellies and jams. They are an excellent fruit to bake in a pie. The leaves are high in vitamin C and can be made into a delicious midsummer iced tea.

B

BORAGE
Borago officinales

Growing
- Hardy annual
- Well-drained soil
- Full sun
- 60 cm high by 40 cm wide

Harvesting
Pick leaves and flowers throughout spring and summer. Flowers can be dried for use in winter.

Parts used
Leaves, stems, seeds and flowers

Uses
- Cooling
- Moistening
- Sweetening
- Cleansing
- Antidepressant
- Adrenal stimulant
- Expels phlegm
- Promotes lactation
- Calms skin conditions

Cautions
Fresh leaves and stems can cause a rash on sensitive skin because of their prickly hairs.

Borage is the ultimate happy herb. As Nicholas Culpeper wrote, borage is helpful 'to comfort the heart and spirits of those that are troubled often with swoonings, or passions of the heart'. It is also a happy plant to grow in the herb garden as it has a positive effect on the plants around it.

In the garden
As it has a long taproot and doesn't transplant happily, sow seeds in situ, from spring through to early summer. Grow borage in well-drained soil in full sun. It will self-seed without becoming too invasive. It is a useful companion plant as it deters leaf-eating insects and attracts beneficial ones. Bees love borage flowers because they are full of nectar and easy to access. It is a good companion for tomatoes and strawberries.

Growing in containers
Use a large container as the plant has a deep root.

Healing properties
This herb has a multitude of healing properties. The leaves and flowers alleviate coughs and colds, stimulate mother's milk and boost people suffering from grief or depression. It lowers fevers, detoxifies the blood and is a good tonic. It stimulates the adrenal gland to release epinephrine, the 'courage' hormone, good for people who are stressed or over-worked. It is also a great hangover remedy! It is a calming, soothing plant for many skin conditions. Borage seed oil is high in gamma-linolenic acid (GLA), one of the essential fatty acids.

How to use
Infusion: (leaves and flowers) For a hangover, as a tonic and to detoxify the system, to treat coughs and colds, to relieve depression and to stimulate breast milk.
Juice: (leaves and flowers) To treat depression, anxiety or grief.
Syrup: (flowers) For depression and as an expectorant for tight chests.
Steam inhalation: (leaves and flowers) To treat dry skin.
Cream: (leaves) For dry, irritated skin or rashes.
Compress: (leaves and flowers) For sprains, bruises and inflammation.
Seed oil: Externally to treat eczema.
Infused oil: (dried leaves) Rub onto dry skin and rashes; use as a chest rub for bronchial conditions; rub onto inflamed joints and sprains.

Culinary uses
Young borage leaves add a fresh cucumber taste to salads and any fruit dishes. The flowers can be used fresh and they can be crystallised for cake decoration or frozen in ice cubes as a fun addition to a cocktail.

B

BUGLE
Ajuga reptans

Growing
- Hardy perennial
- Well-drained, moist soil
- Full sun to semi shade
- 10–30 cm high and spreading

Harvesting
Pick the whole plant when it is in flower and dry for later use.

Parts used
Whole plant

Uses
- Astringent
- All-purpose wound healer

Cautions
Do not take internally unless under guidance of a trained herbalist, as it can be toxic.

This low-growing perennial is a useful herb to grow, as it spreads to form a dense, water-retaining ground cover.

In the garden
Propagate by separating and replanting leafy runners in spring or autumn. In spring and early summer it produces spikes of deep blue flowers, which bees and butterflies love. Although bugle prefers moist, well-drained soil, it can survive periods of drought. If planted in full sun, the leaves are smaller and the flowers are bigger. It doesn't mind semi shade, but the flowers will be smaller and the leaves bigger. It can be invasive if it grows in moist soil, but is quite easily controlled if harvested regularly. It is a lovely plant to grow as a ground cover between paving stones.

Growing in containers
Bugle is happy as a container plant but needs plenty of water.

Healing properties
Bugle has been used for centuries as a mild astringent to stop bleeding and to treat cuts, bruises, sores and gangrene. It has fallen out of favour and is no longer used much by modern herbalists.

How to use
Ointment: To treat external wounds, cuts, sores and bruises.
Infused oil: As for ointment.

BULBINE
Bulbine frutescens

Growing
- Hardy perennial
- Well-drained soil
- Full sun
- 20 cm high by 30 cm wide

Harvesting
Pick leaves whenever required.

Parts used
Leaves

Uses
- Soothing
- Cooling
- All-purpose wound healer

Hardy and drought resistant, bulbine is indigenous to South Africa. There are several varieties but the most common is the one with yellow flowers. Bulbine is sometimes incorrectly called bulbinella, a similar plant, but without the fluffy stamens.

In the garden
This is an ideal plant for a low-maintenance herb garden as it is one of the easiest and most forgiving plants in the world to grow. The quickest way to grow it is to find a friend with a plant and split off a section with roots and replant it in your garden. It spreads happily without becoming invasive and has flowers almost all year round. It is very drought resistant and prefers well-drained soil in full sun. It will grow in semi shade, but then won't produce as many flowers. It is the perfect plant to fill those spaces where nothing else will grow. If it becomes too thick, simply divide the clumps in spring (see page 25) and give plants away to friends.

Growing in containers
Bulbine is ideal as a container plant. Make sure it is not over-watered.

Healing properties
Bulbine has long been used by traditional healers for a variety of ailments and its usage probably dates back to the first inhabitants of our continent. Similar to the aloe, bulbine's fleshy leaves produce a viscous sap that has many uses. It is particularly good for treating skin problems such as eczema, cold sores, itchy skin and rashes, dry skin and insect stings. It helps alleviate itchy rashes on dogs and won't be harmful for them if they lick it off.

How to use
Fresh: Split the leaf and apply the gel directly to the affected area. To treat burns, cold sores, sunburn, insect bites, inflamed and itchy skin (especially good if you are itchy from handling plants), rashes, eczema and shingles.
Cream: Add fresh gel to moisturising creams for dry skin and to ease rashes.

B

BURDOCK
Arctium lappa

Growing
- Hardy biennial
- Well-drained, fertile soil
- Full sun to semi shade
- 2 m high by 1 m wide

Harvesting
Harvest leaves from first-year plants, before they flower. Harvest roots from first-year plants in autumn. Harvest young flower stalks as the flower buds are forming. Harvest seeds once the seed ball is dry. To dry the roots, split them lengthways before drying in a dark well-aired place.

Parts used
Roots, leaves and seeds

Uses
- Seeds: prevent fever, relieve inflammation, antibacterial, lower blood sugar levels
- Roots: detoxifier, diuretic, antifungal, antibiotic, promote sweating, relieve rheumatism
- Leaves: digestive, diuretic, mild laxative

In 1941 a Swiss inventor returned from a hunting trip in the Alps with his dog. As any good inventor would, he became curious when he saw the burdock burrs that were stuck to his clothes and his dog's coat. He took a closer look at the hooks under a microscope and after eight years of experimenting he produced Velcro.

In the garden
Burdock likes well-drained fertile soil in full sun but will also grow in semi shade. It does not like to dry out and, to produce the best roots, it needs regular watering. It is a biennial and produces a large rosette of leaves in its first year. In its second year it produces a long, leafy flower stalk with thistle-like flowers, which ripen into little seed balls with hooks. If you don't want it to seed itself, cut the flower stalks back before the seedpods dry.

Growing in containers
As burdock has a very long root, it is not a suitable plant for growing in containers.

Healing properties
Burdock is a cleansing herb, purifying the blood and clearing the body of built-up toxins, especially those leading to skin problems. It effectively treats eczema, psoriasis, neuralgia, shingles, acne and boils as well as rheumatism, gout and painful arthritis. It helps speed up the digestive system, reduces fevers and lowers blood sugar levels.

How to use
Infusion: (leaves) Take before meals to ease indigestion and as a digestive stimulant and a mild laxative.
Decoction: (roots) To clear skin disorders; particularly good for boils, acne and eczema. Use as a wash for acne and fungal skin conditions such as athlete's foot and ringworm.
Decoction: (seeds) To reduce fevers, and treat a sore throat and cough.
Tincture: (roots) Use to detoxify the body; to treat rheumatism, gout and arthritis; and to stimulate digestion.
Poultice: (roots) Apply to skin conditions.

Culinary uses
The long taproot of young burdock plants is a very popular vegetable in Japan, most often added to miso soup. Steamed or roasted until tender, it has a sweet, earthy flavour.

Other uses
Burdock has very deep roots, making it a good nutrient-accumulating plant in the garden. Its leaves will add nutrients to compost or nearby plants.

Cautions
In its first year burdock looks very similar to rhubarb so don't make the mistake of confusing the two, as rhubarb's leaves are poisonous.

C

CALENDULA
Calendula officinalis

Calendula is sunshine in a blossom. This plant was Cleopatra's secret weapon, as the ancient Egyptians used calendula-infused oil as an anti-ageing remedy.

Growing
- Hardy annual
- Well-drained, fertile soil
- Full sun
- 60 cm high by 60 cm wide

Harvesting
Harvest young leaves and open flowers from late autumn through to summer. The flowers and leaves can be dried for later use.

Parts used
Leaves and flowers

Uses
- Pungent
- Drying
- Cooling
- Astringent
- Antiseptic
- Antifungal
- Relieves inflammation
- Menstrual regulator
- Wound healer
- Bile stimulator

Cautions
Do not take internally when pregnant or breast-feeding. Don't use on deep cuts or puncture wounds as rapidly healing skin might close an infection inside the wound.

In the garden
A member of the daisy family, calendula is a versatile, spicy herb that's easy to grow. It flowers from late autumn, throughout winter into early summer. Grow it from seed in full sun in well-drained soil – it does not like wet feet. Calendula plants like being planted in big groups together and this provides a one-stop shop for bees, which love their flowers. Deadhead flowers to encourage further flowering. It will self-seed if you let it go to seed, but never becomes a problem plant.

Its flowers attract beneficial insects and its leaves deter leaf-eating insects. It is a good companion for winter crops of cabbage, cauliflower, broccoli, broad beans and peas.

Growing in containers
Calendula grows well in a sunny container.

Healing properties
Calendula has a wide range of medicinal uses, from astringent to menstrual regulator. It is primarily used as an anti-inflammatory and as a skin healer for problems such as eczema, cuts, mouth ulcers, insect stings, grazes, slow-healing wounds, dry or inflamed skin, perineal tears after childbirth, minor burns, sunburn and fungal conditions such as athlete's foot, thrush and ringworm. When treating a wound, make sure that it is clean and that there is no infection. Tissue treated with calendula heals very quickly and it is vital that no infection remains underneath the healed skin.

How to use
Infusion: (petals and leaves) As a mouthwash for mouth ulcers or gum infections. Externally as a face wash for pimples and acne. Gargle for a sore throat and tonsillitis. Drink for digestive problems and stomach aches, menopause symptoms, infections, to release nervous tension and to relieve menstrual cramps.
Tincture: (petals and leaves) Use internally to treat tonsillitis, sore throats and swollen glands. Use externally to heal pimples, sores, cuts, ringworm and other fungal infections.
Douche: (petals and leaves) For vaginal thrush and infections.
Ointment: (petals) For varicose veins, sore nipples from breast-feeding, bruises, dry, inflamed or itchy skin, perineal tears, wounds and cuts, eczema, sunburn and scalds.
Compress: (petals) Soak a pad in a hot infusion and apply to varicose veins, sore nipples or eyes, slow-healing wounds or stiff muscles. Soak a pad in a cold infusion for sprains and sore joints.
Essential oil: Add to bath water to relieve depression and nervous tension.
Infused oil: (dried petals) As for essential oil. Apply to varicose veins, aching joints, sore nipples and stiff muscles. Add to face cream for an anti-ageing remedy.

Culinary uses
Known as the 'poor man's saffron', calendula's golden orange flowers colour food yellow and have a spicy, tangy flavour, similar to saffron. Young leaves add a slightly bitter bite to leafy salads. Fresh petals add a yellow tint to soups, pasta or rice dishes, herb butters and salads. Dried petals are more concentrated.

CALIFORNIA POPPY
Eschscholtzia californica

Growing
- Hardy annual
- Well-drained soil
- Full sun
- 60 cm high by 30 cm wide

Harvesting
Harvest leaves, flowers and seeds to use fresh or dry for later use.

Parts used
Whole plant

Uses
- Calming
- Pain reliever
- Sedative

This poppy is native to California and is its state flower. For centuries, Native Americans have used it for its relaxant, mildly sedative and antispasmodic qualities.

In the garden
The gorgeous, golden-orange flowers of California poppy add a splash of sunshine to the garden and bees go crazy for them. The plant grows easily from seed in all types of soil and will self-seed without becoming invasive. Direct seed California poppies in spring or in autumn in full sun. Each flower will produce a long, thin seedpod. If you continually deadhead these, the plant will flower for longer. In midsummer, leave some to go to seed and cut the others back. They will resprout for an autumn flowering. It is a hardy plant and will provide bright colour in the garden during spring and again in autumn and winter.

Growing in containers
California poppies will happily grow in a container. Do not over-water.

Healing properties
This is a relaxing herb with calming and sedative qualities. It is good for treating nervous conditions, especially insomnia. It is also a pain reliever, particularly for toothache.

How to use
Infusion: To prevent insomnia and ease anxiety and nervous tension. As a sedative to ease pain.
Poultice: (fresh root) Apply to sore teeth.

CANCER BUSH
Sutherlandia frutescens

Growing
- Short-lived perennial, will survive mild frost
- Well-drained, fertile soil
- Full sun
- 2 m high by 1 m wide

Harvesting
Harvest leaves to dry in midsummer.

Parts used
Leaves

Uses
- Restorative
- Immune system booster
- Detoxifier
- Antibiotic
- Antifungal
- Relieves inflammation
- Antioxidant
- Reduces stress
- Antiviral

Cautions
Do not exceed 600 mg dose per 70 kg weight of person being treated.

Sutherlandia is one of the most powerful of South Africa's indigenous herbs. It has been used for centuries by traditional healers, and modern-day researchers are now discovering a multitude of medicinal uses for this wonder plant.

In the garden
Apart from its medicinal properties, the cancer bush is a pretty plant to add to the herb garden. It has delicate grey-green leaves and bears vivid scarlet flowers, which are followed by fat green seedpods. It is grown from both seeds and seedlings and likes full sun and warm conditions with fertile, well-drained soil.

Growing in containers
Cancer bush will happily grow in a large container as long as it is watered regularly.

Healing properties
Traditionally, the cancer bush is used for numerous treatments ranging from cleansing wounds to reducing fevers. Modern research has discovered that it contains potent elements that effectively invigorate and strengthen the immune system. It is particularly efficient at helping the body deal with stress and depression and to ward off disease.

How to use
Infusion: (dried leaves) As an external wash to treat burns and skin conditions, and to cleanse wounds. Internally as a restorative tonic and to strengthen the immune system, to reduce inflammation, and to reduce fevers and the symptoms of colds and flu. To detoxify the body and treat liver and urinary tract infections. To help treat HIV/Aids, cancer, asthma and bronchitis and diabetes. It stimulates appetite in recovering patients and speeds up healing.

CAPE GOOSEBERRY
Physallis peruviana

I used to think this plant got its name because it came from the Cape. Then I discovered it was named after the cape-like husk that grows over the fruit. It is a member of the nightshade family of plants, along with tomatoes and potatoes.

Growing
- Perennial, annual in cooler climates
- Well-drained soil
- Full sun to semi shade
- Rambling plant can reach 3–4 m long

Harvesting
The berries ripen from mid to late summer through to late winter in warmer areas. They are ready to eat when they are golden yellow and the husk changes from green to a transparent straw colour. The berries will keep for over a month if they are not removed from their husks.

Parts used
Fruit and leaves

Uses
- High in vitamin C, vitamin A and potassium
- Antioxidant
- Antiviral
- Relieves inflammation

Cautions
Do not eat unripe berries as they are toxic.

In the garden

The cape gooseberry is a rewarding and easy plant to grow. It does well in poor soil and too much fertiliser will result in more leaves than fruit. It is quite a water-wise plant, surviving periods of drought. It prefers well-drained soil and won't do well with muddy, wet feet. Although it is a perennial, it is grown as an annual in colder parts of the country as it dies back in a hard frost. In milder frost areas the leaves will die back but the roots survive. Cut back all the dead branches in early spring and it will pop up again. It is happiest in full sun but doesn't mind semi shade. Sow from seed or transplant seedlings throughout spring into early summer.

This is a rambling plant, which benefits from having space to spread or something to climb over. I have found it does well in amongst shrubby plants, where it climbs up and through them for support and in return the shrubs protect the gooseberry from frost. Birds are fond of the fruit and they can spread seed (and gooseberry plants) all over the garden.

Growing in containers

Cape gooseberry grows happily in a large container and will benefit from some caging or support to keep its growth in check.

Healing properties

The cape gooseberry might be small but it packs a big punch. It is high in potassium, vitamin C and vitamin A. With high levels of beta carotene, it helps relieve hay fever and allergies and also has antioxidant, antiviral and anti-inflammatory properties. In Columbia (where the plant originates) the leaves are used as a diuretic. It has been grown in South Africa for hundreds of years and Zulu healers use the leaves to treat abdominal problems in children.

How to use

Infusion: (fruit) Drink as a cleanser for kidney and liver problems and to reduce fluid retention. To relieve asthma and allergies.
Poultice: (leaves) Heat leaves and apply as a poultice to reduce inflammation.

Culinary uses

Fresh cape gooseberries have a unique tart yet sweet flavour, making them the perfect match for meringues or a good addition to fruit salads. Their tartness mellows with cooking and they are delicious baked as a crumble or pie. They have high pectin content and quickly set when used for preserves or jam. They can also be dried or frozen whole as a nutritious snack.

C

CARAWAY
Carum carvi

Growing
- Hardy biennial
- Well-drained, fertile soil
- Full sun
- 60 cm high by 20 cm wide

Harvesting
During its first year, the leaves can be harvested. In the second year the seeds can be harvested. Wait until the flowers turn brown before cutting them off. Tie a bag around the seed heads and hang them to dry. The seeds will easily shake out once they are ready. Harvest the roots in autumn of the second year.

Parts used
Seeds, leaves and roots

Uses
- Pungent
- Drying
- Warming
- Stimulant
- Relieves spasms

Cautions
Do not use medicinally when pregnant or breast-feeding.

Caraway seeds have been found in archaeological sites dating back 5 000 years. Apart from its healing powers, caraway was believed to have strong retentive powers, hence was used as a love potion to attract and keep lovers. Caraway seeds were sprinkled over prized possessions in the hope that this would keep burglars at bay – however, I don't think I will replace my security company with caraway just yet!

In the garden
Caraway grows easily in almost all climates except for very hot and humid ones. A member of the same family as the carrot, it can be grown for its taproot. It likes well-drained soil and does not transplant well. Sow the seeds in full sun in well-drained, fertile soil. It will die down in winter, but grows back quickly in spring and will soon flower. It self-sows and just needs to be thinned out. The greenish-white flowers attract many beneficial insects, especially bees and parasitic wasps. Do not plant near fennel but do plant near peas.

Growing in containers
Caraway is only suitable to be grown in a deep container.

Healing properties
Caraway seeds are excellent for soothing menstrual cramps, and relaxing and strengthening the muscles of the digestive system. Caraway aids digestion, soothes gas, colic and flatulence and reduces the accumulation of toxins.

How to use
Infusion: (crushed seeds) To aid digestion, soothe cramps and reduce flatulence.

Culinary uses
Young leaves and flowers can be used in salads. Caraway root has a delicate flavour and simple steaming works best. Caraway seeds can be used whole or ground, added to bread and cookies. Caraway seeds go well with bean dishes as they reduce flatulence.

CATNIP
Nepeta cataria

Growing
- Hardy perennial
- A range of soils
- Full sun to semi shade
- 80 cm high and spreading

Harvesting
Harvest the leaves whenever required. The leaves can be dried for later use.

Parts used
Leaves

Uses
- Bitter
- Astringent
- Cooling
- Relieves spasms
- Relieves flatulence
- Digestive stimulant

I sometimes find my 20-year-old cat Roo snuggling amongst the catnip. Despite being completely blind and deaf, she still manages to find her way to her favourite plant. This herb not only keeps cats happy, it is also beneficial to its plant neighbours.

In the garden
Catnip grows easily from seed, seedlings or propagated from cuttings in spring. (See page 23. Make sure you protect newly planted slips from your cats — they can destroy them!) The seeds should be sown directly and shallowly in a moist, sunny position. Catnip is not terribly fussy about its soil and doesn't mind some shade. It can become quite bushy and will spread (as it is part of the mint family). Regular cutting back will maintain its shape and the trimmings are useful as an anti-insect mulch on newly seeded areas. Catnip is reputed to repel rodents and a barrier of catnip around your compost pile is a good idea.

Growing in containers
Catnip grows easily in containers.

Healing properties
Internally it is particularly good for feverish conditions such as colds and flu. It is a calming herb, with its mildly sedative qualities useful for helping insomnia, reducing excitability, palpitations, nervous indigestion and stomach upsets. Externally it is used to treat sore joints.

How to use
Infusion: (leaves) To bring down fevers and as a mild sedative; to treat the digestive system, colic and tense stomachs; to prevent insomnia.
Poultice: (leaves) Apply to sore joints.

Culinary uses
Leaves can be added to drinks, salads and sauces where they add a minty lemon flavour. Add dried leaves to rubs for meat and fish.

Other uses
Cats do love catnip — hence the name. See page 257 for instructions on how to make a catnip mouse using dried leaves and a piece of material.

CHAMOMILE
Anthemis nobilis and *Matricaria recutita*

Growing
- Perennial (Roman) or short-lived annual (German)
- Well-drained, moist soil
- Full sun to partial shade
- Roman: spreading
  German: 40–60 cm high by 20 cm wide

Harvesting
Harvest flowers when they are fully open and dry them as quickly as possible.

Parts used
Flowers

Uses
- Bitter
- Warming
- Moistening
- Relieves inflammation
- Relieves spasms
- Sedative
- Prevents vomiting
- Antifungal
- Antibacterial

Cautions
Avoid chamomile oil during pregnancy.

Chamomile is a gentle, sweetly-scented herb and chamomile tea is well known for its relaxing properties. This is a herb with the reputation of attracting money, so if you are going gambling, wash your hands in a chamomile infusion to ensure winnings!

In the garden
The most common varieties of chamomile are Roman and German. Roman chamomile is a low-growing perennial also known as lawn chamomile. German chamomile is a taller annual with stronger medicinal qualities.

All chamomile likes a sunny spot (but will adapt to partial shade) and light, well-drained, moist soil. Sow seeds in autumn, but do not cover, as they need light to germinate. Keep well watered during dry weather. A chamomile spray will help strengthen plants and combat fungal diseases in the garden. It is especially good to prevent a fungal disease called 'damping off' on young seedlings. It is an all-round good companion plant as it accumulates nutrients and is particularly good near onions and any member of the cabbage family.

Growing in containers
Chamomile grows well in containers.

Healing properties
Chamomile is well known for its mild sedative effect on the nervous system. Taken internally, it helps ease tension, encourages a good night's sleep and eases colic and menstrual cramps. It also assists digestion and relieves nausea. Externally, the soothing and anti-inflammatory properties of chamomile are good for eczema, rheumatism and any inflamed skin condition as well as promoting wound healing. Chamomile is generally a good pain-relieving treatment, especially for those with a low pain threshold or for children.

How to use
Infusion: Drink to treat irritable bowel syndrome, colic and menstrual cramps, to increase appetite, ease tense indigestion, reduce nausea or prevent motion sickness. To ease anxiety, hyperactivity, neuralgia, stress and at night for insomnia.
Mouthwash: To treat mouth ulcers, inflammation and toothache.
Eyewash: Mix 5 to 10 drops in half a cup of water and bathe strained eyes or conjunctivitis.
Steam inhalation: Inhale for asthma, hay fever or bronchial congestion.
Ointment: For insect bites, sunburn, wounds, neuralgia, fungal conditions, eczema and itchy skin.
Essential oil: Add to bath for a calming effect; add to massage oil for rheumatism. As for ointment.
Infused oil: (dried flowers) As for essential oil.

Culinary uses
Add chamomile petals to salads and fruity desserts. Use sugared flowers to decorate cakes.

Other uses
Chamomile spray, made from an infusion of flowers, will help get rid of musty smells in cupboards or clothes. This also works for smelly shoes.

CHICKWEED
Stellaria media

Growing
- Tender, annual
- Any soil
- Full sun to semi shade
- Spreading

Harvesting
Harvest whenever needed. Can be dried for later use.

Parts used
Leaves, flowers and roots

Uses
- Sweetening
- Moistening
- Cooling
- Astringent
- Relieves rheumatism
- Antiseptic

I used to pull this weed out of my garden until I discovered what a useful little plant it is. Herbalist John Gerard wrote in 1597 that chickweed 'comforteth, digesteth, defendeth and suppareth very noticably'.

In the garden
This plant grows effortlessly and will spread. But it has very shallow roots that don't like to be disturbed so is easily controlled. Now I have realised that it is not an unwanted weed but a useful medicinal plant, I don't mind it popping up. If it does become too rampant, pull it out and dry it, or feed it to the chickens – they love it!

Growing in containers
Chickweed will easily grow in a container.

Healing properties
Chickweed is a calming and cooling antiseptic herb, good for many skin ailments. It is a diuretic and tonic, treating rheumatic pains and gout. The root is used to reduce fevers, stop nosebleeds and reduce menstrual bleeding.

How to use
Infusion: (leaves and flowers) To aid digestion; as a cleansing tonic to treat rheumatism and urinary tract infections; and to relieve tiredness. As a wash for eczema and psoriasis. As a hot compress for painful joints.
Decoction: (roots) To reduce fevers, especially in chronic illnesses.
Tincture: (leaves and flowers) Add to rheumatism treatments.
Cream: (leaves and flowers) For eczema, psoriasis and itchy skin conditions. Will draw out infection and toxins from boils, splinters or insect bites.
Poultice: (leaves and flowers) To treat abscesses, boils and painful rheumatic joints.
Infused oil: (leaves and flowers) To treat inflamed skin and rashes or added to the bath to treat eczema.

Culinary uses
Chickweed is high in vitamin C and phosphorous and has a fresh taste. Young leaves can be added to salads or cooked in the same way as spinach.

Other uses
As its name suggests, chickweed makes great food for any birds.

CHICORY
Cichorium intybus

Growing
- Hardy perennial
- Light, alkaline soil
- Full sun
- 1m high by 30cm wide

Harvesting
Harvest young leaves for fresh use. Pick before flowering for drying. Harvest roots in autumn.

Parts used
Roots, leaves and flowers

Uses
- Bitter
- Cooling
- Diuretic
- Laxative
- Tonic

Cautions
Do not use excessively as it can impair the retina.

Chicory is an ancient herb, used for over 5000 years both medicinally and as food for humans and livestock. It has the magical reputation of being able to undo all locks. Don't throw away your keys, though, as this is probably more of a figurative meaning, referring to chicory's ability to clear obstacles in life.

In the garden
Chicory grows easily from seed. Sow in seed trays in early March ready for planting out in late autumn. Chicory likes full sun and fertile, well-drained soil, preferably on the alkaline side. It will go to seed quickly as soon as the weather warms up.

Growing in containers
Chicory is a tall plant with quite a long root so is suited only to a large, deep container.

Healing properties
Similar to dandelion, chicory is full of vitamins and minerals, making it a good tonic herb. Its bitterness increases the flow of bile. It effectively treats gallstones and liver problems. It is a diuretic and cleanses the body of uric acid, making it a good herb for treating gout and rheumatism.

How to use
Infusion: (leaves and flowers) As an uplifting tonic, to treat rheumatism and gout, anaemia, liver problems and weak vision.
Syrup: (roots) As a laxative.

Culinary uses
If you like the bitter taste of chicory then add some young leaves to salads – the older the leaves, the more bitter they become. The flowers are edible and can also be added to salads.

CHILLI
Capsicum frutescens

Growing
- Tender perennial
- Well-drained, fertile soil
- Full sun
- 30–60 cm high by 30–40 cm wide

Harvesting
Harvest when the fruit is a decent size. Nip small ones off with finger and thumb. Use scissors for cutting larger ones to avoid damaging plant. Chillies dry well for later use.

Parts used
Fruit

Uses
- Hot, pungent and drying
- Circulatory stimulant
- Promotes sweating
- Digestive stimulant
- Antiseptic
- Antibacterial
- Increases blood flow to an area

Cautions
Avoid touching the eyes or open wounds after handling chilli. Avoid medicinal use during pregnancy and breast-feeding. Use with caution when applying directly to the skin as it can cause burning.

Although chillies have been used for more than 6000 years in South America, it wasn't until 1490 that they began to spread to the rest of the world. Not just a spicy addition to our kitchens, chillies have valuable medicinal properties too.

In the garden
Chillies prefer a well-drained, fertile soil and love warm climates. They don't like the cold and do not survive a hard frost. Sow seeds in a sunny spot where they are to grow or start indoors in seed trays in late winter. They prefer humid conditions when flowering and setting fruit. Feed them with a high-potassium, organic fertiliser when they begin flowering.

Growing in containers
Chillies grow well in pots; this is a good choice for frost areas, as the pot can be brought indoors in winter.

Healing properties
The variety most often used in medicine is cayenne pepper. This warming plant increases circulation and relaxes spasms and tension. It treats colds, fevers and chills by promoting sweating, and treats shock and depression. It stimulates digestion and increases blood flow to an area, encouraging healing.

How to use
Infusion: For colds, fevers and chills, to increase circulation and warm cold feet and hands. To stimulate digestion and treat shock and depression. As a compress for sore joints.
Tincture: To stimulate digestion and circulation. Diluted as a gargle to treat sore throats.
Ointment: For unbroken chilblains.
Infused oil: To treat neural pain. Mix with a massage oil to treat backache and painful joints. Must be used very diluted, only on small areas and rubbed in well.

Culinary uses
Chillies can be used fresh in stir-fries, curries, stews and soups or made into sauces, chutneys, jellies and jams. Use dry pods whole or grind to a powder for a spice mix or a rub.

CHIVES
Allium schoenoprasum

Growing
- Hardy perennial
- Well-drained soil
- Full sun
- 20–40 cm high, width varies with variety

Harvesting
Don't cut the tops as if you are giving them a hair trim, as this will just weaken the plant. Harvest by snipping off leaves 2 cm above the base with a pair of scissors. They will regrow quickly.

Parts used
Leaves and flowers

Uses
- Mild antiseptic
- Digestive
- Expels phlegm

Much easier to grow than onions, delicious chives are an easy perennial that should be grown in every herb garden.

In the garden
There are various types of chives, from skinny leafed onion-flavoured ones to flat-leafed, garlic types. The flowers, as with other members of the allium family, are pom-pom shaped, from white through to pinky mauve. Chives are easily grown from seed or seedlings. They prefer full sun and moist soil but can handle dry weather. They do a great job of protecting vegetables from many harmful insects. Every couple of years, divide chives in spring by lifting them and splitting them into new plants (see page 25).

Growing in containers
Chives grow easily in containers.

Healing properties
As with all members of the allium family, chives contain sulphur compounds and many essential minerals, which ease digestion and help the body ward off colds.

How to use
Infusion: (leaves and flowers) As a digestive aid. To prevent and treat colds.

Culinary uses
Chive flowers can be tossed onto a salad or added to ice cubes and dropped into a cold soup. Chopped fresh leaves are great on just about anything. They are also delicious used to flavour oil, vinegar or salt.

C

CLOVER
Trifolium (repens, pratense, incarnatum)

Clover leaves have long been associated with luck and prosperity; a belief dating back to the times of ancient Druids. This is a healthy herb for both our garden and us. There are various types of clover, from the most common white-flowered variety (*repens*) to the pinkish-red flower of red clover (*pratense*) and the annual crimson clover (*incarnatum*).

Growing
- Hardy, drought-resistant perennial and annual
- Well-drained soil
- Full sun
- 20 cm high and spreading

Harvesting
Pick flower heads, with upper leaves, in summer, when the flowers are fully open. Clover dries well for later use.

Parts used
Flowers and leaves

Uses
- Sweetening
- Cooling
- Antiviral
- Antifungal
- Oestrogen-like effects
- Relieves spasms
- Expels phlegm
- Anticancer blood purifier

In the garden
Clover, a nutrient-accumulating plant and legume, is an enriching, quick-growing green manure in the garden. Crimson clover is an annual and is a good crop to enrich the soil during winter. To add the nutrients to the soil, cut it back before it flowers. All clovers prefer light, well-drained soil in full sun, but don't mind some afternoon shade. Sow seeds in situ and it will quickly germinate and thickly cover the ground. If you want to use it medicinally, leave it to flower. Bees love the flowers, particularly the fragrant flowers of white clover.

Growing in containers
Clover will happily grow in a deep container.

Healing properties
Clover is high in vitamins, proteins and minerals. Although white clover does have healing properties, red clover is the variety mostly used for medicinal purposes. It is a liver detoxifier and blood cleanser. High in phytoestrogens, it eases the symptoms of menopause and helps treat cancers of the breast, ovaries and lymphatic system. It is an expectorant, loosening tight chests and relieving bronchitis. It heals skin conditions such as psoriasis and soothes burns and sores.

How to use
Fresh: (flowers) Rub on stings and insect bites.
Infusion: (flowers and leaves) Drink to cleanse body when suffering from liver problems, boils, ulcers or abscesses.
Tincture: (flowers and leaves) Internally to cleanse liver, for eczema and psoriasis, menopause and cancer.
Syrup: (flowers and leaves) As an expectorant and to soothe coughs.
Ointment: (flowers and leaves) For skin conditions, burns and swollen lymph glands.

Culinary uses
Add clover flowers to salads or freeze in ice cubes for drinks. Add a handful of chopped clover flowers to muffin and bread mixes.

Red clover's Latin name, pratense, *is from* pratensis, *meaning 'growing or found in meadows'*

C

COMFREY
Symphytum officinale

Growing
- Hardy perennial
- Moist soil
- Full sun to semi shade
- 30 cm high by 60 cm wide and spreading

Harvesting
Pick leaves from late spring onwards. Pick flowers in early summer and dig roots from plants that are two years old or older. For drying, gather leaves before the plant begins flowering.

Parts used
Leaves, stems, roots and flowers

Uses
- Heals bones and connective tissue
- Heals wounds

Comfrey is one of the ancient aristocrats of the herb garden; it is reputed to have been one of the plants in the Garden of Eden. Comfrey should be grown in every organic vegetable garden, as it makes one of the best natural fertilisers.

In the garden
Comfrey grows easily from cuttings (see page 23) or from seedlings. Large plants can be divided in early summer to create new ones. It likes full sun but can handle semi shade and is not too fussy about its soil. It has a very deep root system and will improve poor soil around it. Comfrey leaves contain two to three times more potassium than kraal manure, as well as high levels of phosphorous, calcium, nitrogen and other nutrients. Using its deep roots, comfrey mines these from the subsoil beyond the reach of other vegetables.

    Comfrey does take up quite a lot of space in the garden as it grows into a bushy plant but it is worth having, even in a small garden, because it is so useful. As it grows bushier, keep it trimmed by cutting off leaves. These can be added to the compost heap, where they help it to break down extremely quickly, or shredded and used as mulch (this is excellent for tomatoes). Leaves can also be used to line planting holes (particularly good for potatoes). Let them wilt for a few days and then line the edges and bottom of the hole before planting. They will decompose and supply the plant with rich fertiliser. (Avoid using flowering stems as these can root themselves.) The flowers, which appear from the middle of spring onwards, attract bees to the garden.

Although comfrey has the reputation of being an invasive plant, I have had one comfrey plant in my vegetable garden for years and have no problems with it invading or popping up anywhere unwanted. I grow mine in a corner edged by stones, which prevents it spreading uncontrollably. If comfrey does grow too big, divide it and compost the leaves. It dies back in winter in colder areas but will pop back up again in spring.

Growing in containers
Comfrey can grow in a deep container.

Healing properties
Comfrey's Latin name *symphytum* comes from the Greek *sympho* meaning 'to unite'. Comfrey is from the Latin *confirmare* meaning 'to join together'. This gives a good idea of comfrey's healing abilities. Culpeper wildly claimed that 'the roots being applied outwardly, cure fresh wounds or cuts immediately, being bruised and laid thereto; and is specially good for ruptures and broken bones, so powerful to consolidate and knit together that if they be boiled with dissevered pieces of flesh in a pot, it will join them together again.' Although comfrey is not as magical a healer as Culpeper believed, it does contain allantoin, a compound that speeds cell renewal in muscles, connective tissue and bones. For over 2000 years herbalists have used 'knitbone', as it was commonly called, to speed the healing of broken bones, fractures, bruises and burns. It is also effective for slow-healing wounds, painful joints and various skin conditions.

How to use
Infusion: (leaves) Add to your bath (keep your heart above the water level) and soak painful joints for 20 minutes.
Tincture: (roots) Rub onto painful joints.
Cream: (flowers, leaves and roots) Apply externally to any longer-term muscle or bone damage, such as osteoarthritis. Apply to slow-healing wounds, burns, dry skin and varicose veins. Will reduce scarring and prevent stretch marks.
Poultice: (flowers and leaves) Purée into a paste. Apply to sprained limbs, torn cartilage, small hernias, torn ligaments or fractured bones that are difficult to set in plaster, such as ribs, toes or hairline cracks in larger bones.
Poultice: (roots) Mash into a paste and apply to small broken bones.
Infused oil: (dried flowers, leaves and roots). Use hot infusion method. Apply to arthritic joints, bruises, sprains and strains.

Other uses
Comfrey leaves make a rich, fermented tea for plants. See page 258 for the recipe.

Cautions
Although it has been taken internally for centuries, recent studies show that comfrey contains potentially toxic alkaloids, so avoid internal use. Be careful when working with comfrey as its leaves are prickly and can cause a rash. Make sure wounds are well cleaned before applying comfrey, as rapidly healing skin might close an infection inside the wound.

C

CORIANDER
Coriandrum sativum

Growing
- Hardy annual
- Well-drained soil
- Full sun, some afternoon shade
- Goes to seed quickly in hot weather
- 45 cm high by 30 cm wide

Harvesting
Harvest leaves as required, ensuring enough are left for the plant to keep growing healthily. Harvest seed heads as they ripen.

Parts used
Whole plant

Uses
- Warming
- Detoxifying
- Aids digestion
- Pain reliever
- Stimulates appetite and libido
- Reduces cholesterol

Cautions
Do not use coriander in more than culinary amounts when pregnant or breast-feeding.

When I first tasted coriander leaves I thought they tasted like stinkbugs. I then discovered that coriander's Latin name is derived from the word *koris*, meaning bug, so my taste buds knew what they were on about. Luckily my mouth changed its mind and now I love coriander. Also known as cilantro, this is an ancient herb, with Sanskrit texts dating back 7 000 years referring to the cultivation of coriander.

In the garden
It is an easy plant to grow, but does go to seed easily. If you eat coriander regularly, then do successive sowings. Directly seeded coriander does better, as transplanting often causes the plant to bolt. Coriander likes well-drained soil and a sunny spot, but appreciates some shade from hot afternoon sun, especially during hot, midsummer months.

Coriander flowers attract many beneficial insects, both pollinators and predators, and are particularly good planted near cabbages, eggplants and tomatoes. Fennel and coriander will not do well together. If you are growing coriander for the leaves and not the seeds, encourage plants to become bushier by snipping off any flower heads.

Growing in containers
Coriander has quite a long taproot, so choose a deep container.

Healing properties
This warming herb is used to increase circulation, stimulate libido and appetite and relieve arthritis, rheumatism and stiffness. It aids digestion, relieves irritated skin and is a tonic for the nervous system. It also has pain-relieving properties, useful for muscular pain, headaches and neuralgia. It is a diuretic and has antifungal and antibacterial properties, making it a good detoxifier. The seeds lower cholesterol.

How to use
Infusion: (seeds) To sweeten breath, heal mouth ulcers, aid digestion and ease indigestion; to detoxify the liver; to ease nervous tension and ease neuralgia.
Infusion: (leaves and seeds) To stimulate the appetite and libido, increase circulation, reduce cholesterol and aid digestion.
Decoction: (seeds) To sweeten breath.
Cream: (leaves and seeds) To soothe irritated skin and treat fungal conditions.
Compress: (seeds) To ease stiff and painful joints and to ease headaches and neuralgia.
Essential oil: In an oil burner for headaches and nervous tension.
Infused oil: (leaves and seeds) As for essential oil. Apply to rashes and irritated skin; rub on painful joints.

Culinary uses
The seeds, flowers, leaves, stems and roots are all edible. Coriander leaves and seeds taste quite different and are essential elements of Asian, Latin and Indian cuisines. Leaves and roots are used in many curry pastes. The roots and stems can be diced and added to stir-fries. Leaves are added right at the end of cooking or as a garnish. Seeds are used in spice mixes for curries and stews and as flavouring for pickles, vinegars, chutneys and sauces. It is best to store seeds whole and grind them as needed. Roasting whole coriander seeds before grinding adds to their flavour. Fresh green seeds are also delicious – try blending them into curry pastes or use them in stir-fries.

C

CORNFLOWER
Centaurea cyanus

Believed to repel evil, protect property and enable clearer vision through the third eye, the attractive blue cornflower has been associated with folklore, magic, pagan rituals and legends dating back centuries.

Growing
- Hardy annual
- Well-drained soil
- Full sun
- 30–80 cm high by 15 cm wide

Harvesting
Pick flowers when they are fully open. They dry well for later use.

Parts used
Flowers

Uses
- Relieves inflammation
- Stimulant
- Tonic

In the garden
These magical flowers are a wonderful addition to the garden, attracting a host of beneficial insects with their flowers. Although there are pink and purple hybrid varieties, it is the vivid blue ones that steal the show. Sow seeds directly into the soil in a sunny spot in spring and again in late summer for a gorgeous winter and spring flowering. They will self-seed without ever becoming a problem.

Growing in containers
Cornflower plants will be happy in a large container.

Healing properties
Medicinally, cornflowers have traditionally been used as an eye treatment. They are also a tonic and stimulant.

How to use
Infusion: As an eye-strengthening and soothing eyewash. Internally as a tonic and stimulant.

Culinary uses
The flowers have a slightly sweet to spicy, clove-like flavour and can be used for cake decoration and as a wonderfully bright addition to salads.

Other uses
The flowers can be used to make fabric dye or food colouring.

The centaur, the Greek mythological half man half horse, used cornflowers to heal his battle wounds

D

DANDELION
Taraxacum officinale

Growing
- Hardy perennial
- Any soil
- Full sun to partial shade
- 30 cm high by 20 cm wide

Harvesting
Pick young flowers and fresh leaves any time. Collect leaves for drying when the plant is flowering. In spring dig up roots of two-year-old plants and use fresh or dried.

Parts used
Leaves, stem, flowers and roots

Uses
- Cooling
- Bitter
- Diuretic
- Liver cleanser and stimulator
- Skin cleanser
- Tonic

Cautions
Dandelion is a strong diuretic. If you are on medication its efficacy may be reduced.

Dandelion is an incredibly adaptable plant, growing in almost every environment, from fertile soil to rocky slopes and from sea level to miles above the ocean.

In the garden
To most people the dandelion is a weed but I like having dandelion plants in my garden and they have never become prolific enough to be a problem. If you keep an eye on them and pull them out before they go to seed, you can control them. These are beneficial plants in a herb garden. They have very long taproots and, instead of competing with their herbal neighbours, they benefit them by drawing out nutrients and minerals (particularly calcium) from lower levels. When they die, their long root channels provide 'mine shafts' for earthworms, enabling them to penetrate deeper into the soil than they would normally. Dandelions also give off ethylene, which causes fruit and flowers to mature early. But beware: too much ethylene can stunt the growth of neighbouring plants, so it is best to limit your dandelion population.

Growing in containers
Dandelion plants have a long taproot and need a deep container.

Healing properties
Dandelion is known as an alterative, which is a herb that restores health by purifying the body. Dandelion leaves contain high levels of potassium, higher levels of vitamin A than carrots and higher levels of protein, calcium, iron and nearly all vitamins and minerals than lettuce and many other greens. Dandelion has been used for centuries as an effective detoxifier and cleanser. Its high potassium content makes it an excellent diuretic, because it replaces the potassium that is lost. Use it to treat urinary infections, water retention and loss of appetite, and to cleanse the liver and gall bladder. The root and leaf treat breast diseases, joint pain, fever and skin ailments. The next time you find one in your garden, think again about just pulling it out!

How to use
Infusion: (leaves, flowers and stem) Externally and internally to reduce acne, eczema and other skin conditions. Internally as a diuretic to cleanse and stimulate the liver, cleanse the gall bladder, reduce bloating and to treat urinary infections and gout.
Decoction: (roots) For liver problems and to help the body deal with strong chemical drugs. Lowers cholesterol if taken regularly.
Tincture: (root) To detoxify the liver, to stimulate appetite and to aid digestion.
Tincture: (leaves) Added to remedies for people with ailing hearts to ensure adequate intake of potassium.
Juice: (leaves) As a diuretic, for the same conditions as treated by an infusion, but stronger.

Culinary uses
The leaves and flowers of the dandelion are edible and very good for you. Dandelion buds are tastier than the flowers and it is best to pick them when they are tightly bunched in the centre. Eat them raw in salads, steam them or fry them in butter. The sweet, honey-tasting flowers are best when picked young, as mature flowers are quite bitter. They can be added to salads and sprinkled on rice. For years, dandelion flowers have been used to make wine. Young dandelion leaves are a nutritious addition to leafy salads or to stir-fries.

Picking the puffball dandelion seed head, and blowing the little parachutes into the wind, takes me straight back to childhood. We would try to tell the time by how many puffs it took to blow all the seeds off

D

DILL
Anethum graveolens

Growing
- Hardy annual
- Well-drained, moist soil
- Full sun
- 70–80 cm high by 30 cm wide

Harvesting
Start harvesting leaves once the plant is well established. Then gather small amounts from a number of plants. The leaves can be dried but it is best to do it quickly, otherwise they lose their flavour. Harvest seeds once they turn brown (but before they scatter).

Parts used
Seeds, leaves and flowers

Uses
- Aids digestion
- Reduces stress
- Nerve tonic

Dill has been cultivated for over 5 000 years. A herb with a history as long as this is bound to have gathered a reputation and it is said that if you serve a witch a cup of dill tea, it will rob her of her power. Perhaps the reason lies in dill's abilities as a soothing herb.

In the garden
Dill, another member of the parsley family, is easy to grow. It likes full sun, but doesn't mind afternoon shade. Because it doesn't like having its long roots disturbed, sow seed directly where it is to grow. Sow the seeds quite closely together as the plants will then support one another. When mature, they will be about a metre tall, so position them where they won't block other plants' sun. Successive sowings ensure a continuous harvest. Cut the leaves back often to prevent it flowering.

Dill self-seeds if left to flower and the bright yellow umbels attract beneficial insects, especially ladybirds, hoverflies and bees. It is beneficial to potatoes and all vegetables in the cabbage family. Cucumber, lettuce and onion also grow well together with dill. Neither carrots nor caraway like being planted near dill. Don't grow fennel and dill near each other, as they will cross-pollinate.

Growing in containers
Dill can be grown in a large container.

Healing properties
The name 'dill' comes from an old Nordic word *dilla* meaning 'to soothe'. It eases indigestion, flatulence and colic. Use the dried seeds to treat gastrointestinal problems. Particularly good for infants and children, it also improves the flow of breast milk.

How to use
Infusion: (seeds) To treat colic; to increase breast milk.
Infusion: (leaves and seeds) To treat indigestion, flatulence, colic and stomach cramps. To ease restlessness, stress and insomnia.
Tincture: (seeds) As for infusion of leaves and seeds, but stronger.
Dried seeds: Chew for bad breath. To soothe colic and stomach cramps, mash two teaspoons of seeds and add to a cup of boiling water.

Culinary uses
Fresh dill leaves add flavour to salads, egg dishes, herb butters, dips, vinegars and particularly fish dishes. Dill seed has a sharper flavour than the leaves. Use dried seed in soups and vegetable dishes, sprinkled on top of seed bread and as a pickling spice. Dill flowers look great in ice cubes or tossed on top of a green salad.

According to folklore, a sprig of dill hung over the door will protect hearth and home

E

ECHINACEA
Echinacea purpurea

Also known as coneflower, this is a delightful plant to grow in the garden; bees and butterflies love it and its striking pink flowers last throughout summer. Even when it loses its flowers, the cone-like seed heads create an eye-catching display.

Growing
- Hardy perennial
- Well-drained soil
- Full sun
- 70 cm – 1 m high by 50 cm wide and spreading

Harvesting
Harvest flowers and leaves when required. Harvest roots from a mature plant, at least three years old.

Parts used
Leaves, flowers and roots

Uses
- Cooling
- Drying
- Antiseptic
- Antibiotic
- Stimulates the immune system

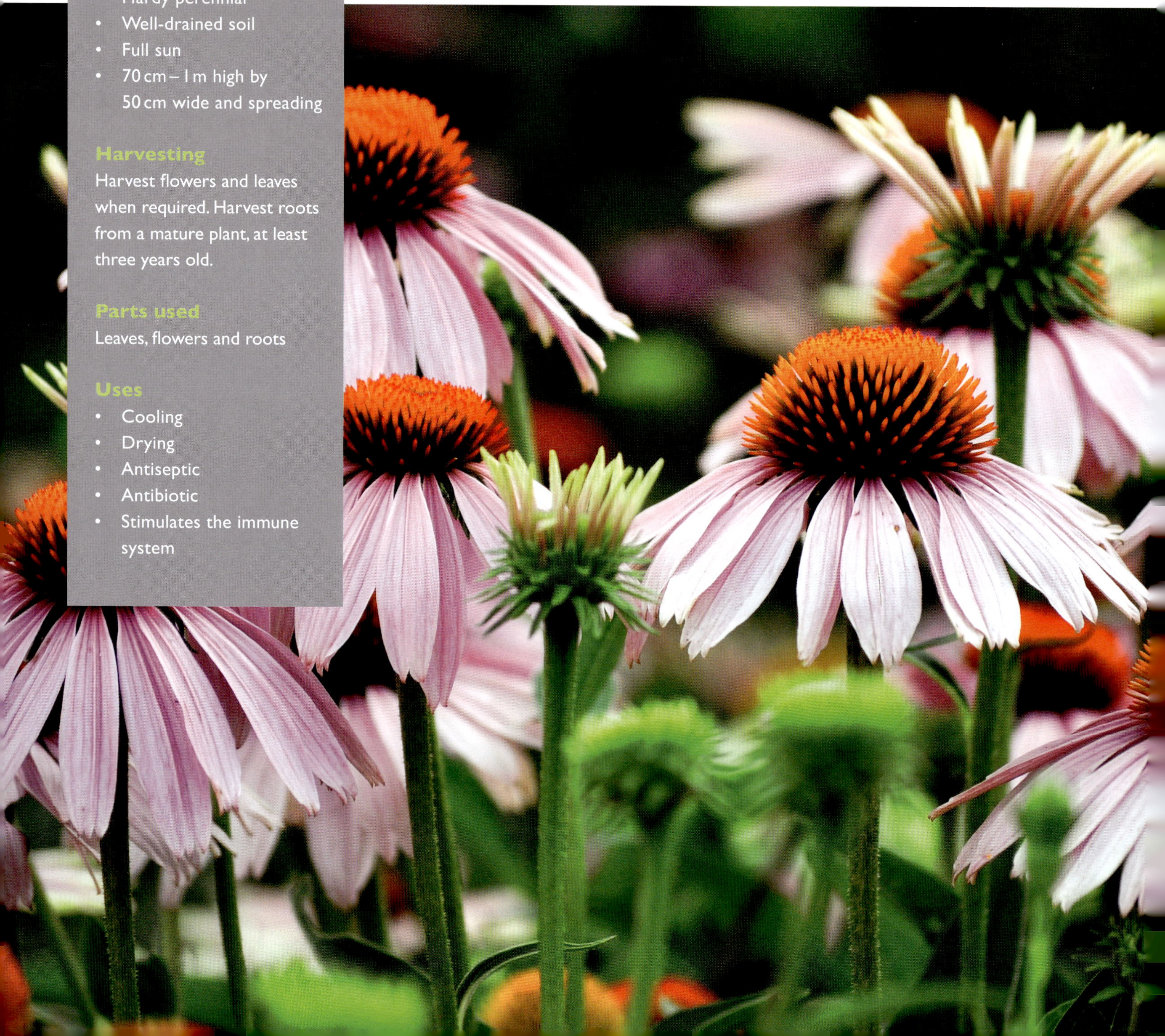

In the garden
Echinacea likes full sun and well-drained soil. It is easily grown from seed and will self-seed happily once established. The plants die back in winter but soon spring up again when the weather warms up. They can also be propagated by division (see page 25). Bigger plants can be divided in late winter and planted out in spring.

Growing in containers
Echinacea will grow well in a large, deep container.

Healing properties
Echinacea is indigenous to North America, where Native Americans have used it for hundreds of years to treat infections and wounds, and as a general 'cure-all'. Today it is widely cultivated, as it contains polysaccharides. These enhance the immune system, relieve pain, lessen inflammation and have antiviral, hormonal and antioxidant properties. The root contains the bulk of the medicinal properties, although the leaves and flowers are also effective. It is an effective immune booster, and a good remedy for hay fever.

How to use
Decoction: (flowers, leaves and root) Drink to stimulate the immune system. To reduce symptoms of hay fever, colds and flu, and infections in general. Drink regularly to reduce allergies. Gargle or use as a mouthwash for sore throats, gum diseases and mouth infections. Externally as a wash for minor wounds, acne, eczema and burns.
Tincture: (root) Apply to insect bites and stings to relieve itching and swelling.

Culinary uses
Use fresh echinacea petals in salads or desserts. Freeze in ice cubes and add to drinks.

Other uses
Research is being done on echinacea's insecticidal properties, as it contains echinolone, which disrupts insect development.

E

ELDER
Sambucus nigra

Growing
- Hardy, deciduous shrubby tree
- Well-drained soil
- Full sun
- 3–4 m high by 4 m wide

Harvesting
Harvest flowers from spring through to summer. Harvest berries when they ripen in late summer to autumn. Both dry well for later use.

Parts used
Flowers, berries (cooked) and leaves (externally only)

Uses
- Warming
- Drying
- Diuretic
- Antiviral
- Relieves inflammation
- Laxative
- Astringent

Cautions
The berries should not be eaten raw. The leaves and bark are mildly toxic and the roots are extremely poisonous.

The elder is associated with many legends and ancient magic. It is believed to protect and bring luck as well as clear negativity, sorcery and evil. However, throughout history, it has also been associated with death. In English funerals, elder branches were trimmed into a cross and planted on graves. If the plant blossomed, it meant the dead were happy. Even the hearse driver's whip was made from elder wood. These beliefs relating to both life and death make sense when you realise that some parts of the elder are poisonous while others are deliciously edible.

In the garden
Elder is not really suited to small gardens as it is quite an unruly, rambling bush. If you do have space, it is a good addition and can be kept trimmed. Either propagate from cuttings (see page 23) or buy a young tree. It likes full sun and well-drained soil. It is a helpful tree to plant near a compost heap, as it has wide-spreading roots which assist in breaking down the compost. The soil underneath and near an elder will be friable and loose. Amongst other things, I use the area around my elder to grow midsummer lettuces, with the bushy elder providing just the right amount of shade and its roots keeping the soil moist.

Growing in containers
Elder is not suitable for containers.

Healing properties
The elder's berries and blossoms have been used for centuries for everything from cough syrups to cosmetics. It is a particularly good plant for colds and coughs. The flowers have decongestant and antiviral properties and can be used to prevent a cold as well as for treating one. The berries are rich in vitamin C.

How to use
Infusion: (flowers) Internally for colds, sinusitis and hay fever, and to reduce fevers. Gargle to treat a sore throat. Externally as a wash for mouth ulcers and tired or infected eyes.
Decoction: (berries) As a cough medicine and a gentle laxative.
Tincture: (flowers) To prevent colds and other viral infections, to treat colds and as a bronchial and nasal decongestant.
Syrup: (berries) To treat coughs and colds and as a mild laxative.
Cream: (flowers) For itchy, dry skin and chilblains.
Infused oil: (leaves with flax oil) Strain the mixture after the oil has absorbed the green colour of the leaves and apply the oil to haemorrhoids.

Culinary uses
The berries are very high in vitamin C and other nutrients, however they are toxic when eaten raw and are best used cooked in jellies, jams, syrups and pies. The flowers can be added to jellies, used to make deliciously refreshing cordial or added to salads. You lose a lot of flavour by washing elder flowers, but they often have little pollen bugs on them. Place the flowers in a colander with a dishcloth over the top and the bugs will quickly crawl off.

Other uses
Use elder leaves as insect-repelling mulch or to make an effective spray to repel insects. (See page 259 for a recipe.) The spray helps repel fleas. Get rid of ants by pouring a strong decoction down their holes.

E

EVENING PRIMROSE
Oenothera biennis

Tall evening primrose, with its bright yellow, sweetly scented flowers, is a must-have in the herb garden. For centuries it was grown as a culinary rather than medicinal herb. However, research in the 1980s discovered that the oil from the seeds has valuable medicinal properties.

Growing
- Hardy, biennial, drought tolerant
- Well-drained soil
- Full sun
- 1.2 m high by 60 cm wide

Harvesting
Pick flowers and leaves in summer. Seeds ripen from midsummer onwards. Dig up roots in the second year, before the seeds start to form.

Parts used
Leaves, flowers, stem, seeds and root

Uses
- Relieves stress and depression
- Improves circulation
- Balances hormones
- Lowers cholesterol
- Astringent

Cautions
Evening primrose should not be taken by epileptics or if on medicine for schizophrenia. Do not use if you are pregnant or breast-feeding.

In the garden
Sow seed directly where it is to grow in spring to early summer. It likes full sun and well-drained soil. It can grow in poor soils, but dislikes too much moisture or soggy soil. In the right conditions it will self-seed quite rampantly if allowed.

Growing in containers
Grow evening primrose in a deepish container, with support, as it grows over a metre tall.

Healing properties
The seed oil is where evening primrose's main medicinal benefits lie. Evening primrose oil contains gamma-linolenic acid (GLA). This is an essential fatty acid that our bodies cannot produce. Evening primrose helps keep skin healthy, balance female hormones, reduce eczema and improve circulation. It also helps reduce swelling and eases depression and hyperactivity.

How to use
Infusion: (leaves and flowers) For asthma and spasmodic coughs.
Decoction: (root) To help with weight loss.
Steaming: (infusion of flowers, leaf and stem) Use as a facial steam to improve the condition of the skin.
Poultice: (leaf) To heal bruises, speed healing of wounds and reduce swelling in painful joints.
Oil: (seed) Internally to strengthen the heart in stressed situations, to relieve premenstrual or menopausal symptoms, to calm hyperactivity and lift depression, to improve circulation, to treat a swollen prostate gland and swollen and painful joints. Externally for eczema, psoriasis and acne; rashes; dry, itchy skin; reducing wrinkles and scarring.
Infused oil: (seeds) As for oil.

Culinary uses
Add fresh evening primrose flowers to salads. Young leaves can be eaten fresh in salads and mature leaves steamed or stir-fried. The roots can be used in soups or roasted.

F

FENNEL
Foeniculum vulgare

During medieval times it was believed that fennel would banish evil spirits. Fennel seeds were thrown at weddings rather than rice, and fronds were pushed into keyholes to prevent ghosts gaining access. This herb should not be confused with its relative, Florence fennel, which produces a bulb and is grown as a vegetable.

Growing
- Hardy perennial
- Well-drained soil
- Full sun
- 1.5–2 m high by 45 cm wide

Harvesting
Cut the feathery fronds and stems as needed. When the flowers turn into seed heads, collect them before they scatter. Harvest the root in late summer to autumn from second-year plants, before the seed has set.

Parts used
Leaves, flowers, stems, roots and seeds

Uses
- Warming
- Drying
- Detoxifier and cleanser
- Relieves flatulence
- Stimulates circulation
- Relieves inflammation
- Diuretic
- Increases breast milk
- Reduces appetite

Cautions
Avoid high doses in pregnancy, as it is a uterine stimulant.

In the garden

Fennel is easy to grow and can be direct sown. It likes full sun and well-drained, fertile soil. It easily grows over a metre tall so be careful of planting it where it will shade other plants. It is a member of the parsley family and has yellow flowers that ladybirds and other beneficial insects adore. Try growing bronze fennel; it tastes much the same but adds colour and interest to the garden.

Beans and tomatoes don't like being planted near fennel. Keep fennel away from dill and coriander, as they can easily cross-pollinate.

Although fennel has a reputation of limiting the growth of plants near it, I have not found this to be the case. The only disadvantage to fennel is that it does tend to self-seed very easily – often where you don't want it. However, if you learn to recognise young seedlings, they are easily pulled when babies.

Growing in containers

Fennel grows well in containers.

Healing properties

It is the seeds and roots that are used medicinally. The seeds help with digestive problems, clear infections in the mouth and eyes, ease coughs, promote the flow of breast milk and relieve colic in babies. They are also a good cleanser for the spleen and kidneys, as is the root.

How to use

Infusion: (seeds) Drink before meals to reduce appetite, drink after meals to aid digestion and treat flatulence, colic and other gastrointestinal problems. Drink to increase breast milk or to relieve colic in infants. Gargle for sore throats or laryngitis. Use as a mouthwash for gum infections and a wash for sore and inflamed eyes.
Decoction: (seeds) For stubborn coughs, abdominal pain and other gastrointestinal problems.
Decoction: (root) For urinary problems and to cleanse the system.
Tincture: (seeds) For digestive problems.
Essential oil: As a chest rub combined with other essential oils.
Infused oil: (leaves and seeds) As for essential oil.

Culinary uses

Fennel leaves are delicious added to salads, soups and stews. Use the stems as a bed for roast chicken or fish, or on the braai to add flavour to meat. The stems can be used as skewers for quick-cooking kebabs. Use the seeds in baking or ground up in curry pastes, sauces and rubs.

F

FEVERFEW
Tanacetum parthenium

With attractive white flowers and insect-repelling properties, feverfew is a useful plant in an organic garden. It has a magical reputation of protecting against illness and injury – particularly effective for people who are accident prone.

Growing
- Hardy, drought-tolerant herbaceous perennial
- Well-drained soil
- Full sun
- 1 m high by 45 cm wide

Harvesting
Pick leaves when needed. Pick flowers as soon as they have fully opened.

Parts used
Leaves and flowers

Uses
- Warming
- Drying
- Bitter
- Relieves inflammation
- Detoxifier
- Digestive stimulant
- Headache reliever

Cautions
Do not take while pregnant or breast-feeding. Do not give to children under two years old. The fresh leaf can cause mouth ulcers in some people. If this happens, eat them with some bread or use the tincture instead. Do not take if you are on blood-thinning medication.

In the garden
Feverfew grows easily from seed and will self-seed in the places it grows best. It likes well-drained soil in full sun. Cut the dead flowers off often to make sure it keeps flowering. It dies back over winter in colder areas but soon springs up in warmer weather. I have plenty of feverfew plants scattered around my garden, as they are useful for various purposes. Its pungent leaves repel leaf- and plant-eating bugs. Sprinkle handfuls of torn-up leaves over newly seeded areas to prevent bugs eating vulnerable seedlings. This also works among lettuces, cabbages and cauliflowers, keeping slugs, snails and leaf-eating caterpillars at bay. It not only repels the insects we don't want – it attracts the ones we do. Its white-and-yellow flowers invite bees, butterflies and hoverflies into our gardens.

Growing in containers
Feverfew grows easily in containers.

Healing properties
Feverfew, a member of the daisy family, has been used medicinally for thousands of years. As its name suggests, one of its uses is to reduce fevers, although it is more commonly used today to treat migraines (especially combined with the bark of the white willow) and headaches associated with menstruation. It limits the inflammation of blood vessels in the head, relieving the pain. It is also used to treat rheumatism and arthritis.

Its bitter taste stimulates the liver, enhancing appetite and digestion and helping clear the body of toxins. It also relieves nausea and vomiting. Its anticoagulant, anti-inflammatory and carminative properties have been used to treat stomachaches, toothaches, insect bites, skin conditions, menstrual problems and dizziness. Research is being done to test feverfew's effectiveness in treating leukaemia.

How to use
Fresh: (leaves) Eat one leaf a day to ward off headaches and migraines. Eat two to three leaves at the first sign of a headache or migraine.
Infusion: (leaves and flowers) Drink a weak infusion to relieve period pains, detoxify the system, stimulate the liver, ease dizziness, nausea and vomiting and stimulate digestion.
Tincture: (leaves and flowers) Take 5–10 drops at the onset of a migraine or headache and every 30 minutes until it eases. For rheumatism and arthritis drink up to 2 ml, three times a day.
Cream: (leaves and flowers) To treat inflamed and dry skin.

Uses
Feverfew contains pyrethrins, which are useful as an organic insecticide but are not as strong as a true pyrethrum plant. Pick open flower heads and dry them on sheets of paper in direct sunlight – the quicker the drying process, the higher the pyrethrin content. Once dried, finely grind the flowers using a mortar and pestle, and sprinkle the powder over your plants. The flowers can also be made into a spray. (See page 259.) Feverfew will keep wool-eating moths out of your cupboards. (See page 213 for a moth-repellent mix recipe.)

F

FLAX
Linum usitatissimum

Growing
- Hardy annual
- Well-drained to sandy soil
- Full sun
- 60cm high by 10cm wide

Harvesting
Pick flowers as soon as they are fully open. Cut the seed heads for drying as soon as they begin to ripen.

Parts used
Flowers and seeds

Uses
- Moistening
- Warming
- Sweetening
- Mucilage
- Contains alpha-linolenic acid
- Relieves inflammation
- Soothing
- Laxative
- Antiseptic
- Expels phlegm

Cautions
Do not use flax seed if you are pregnant or breast-feeding. Do not mistake artists' linseed oil for medicinal linseed oil. Artists' linseed oil is not to be used medicinally.

Flax is one of the world's oldest domesticated crops. Despite being a delicate-looking plant, it has strong fibres that have been used for over 5000 years to make rope, fishing nets and cloth. Oil from the seed, more commonly known as linseed, has been used for centuries, both as a medicine and for preserving wood, waterproofing, lubricating and painting.

In the garden
Flax is a quick-growing annual with gorgeous blue, edible flowers. It is a good addition to a mixed planting. Easily grown from seed, it will self-seed if left to flower, although it never becomes a nuisance. It needs full sun and well-drained to sandy soil. With its shallow root system, it is a particularly good companion for carrots and potatoes, and its flowers attract pollinating insects. It is best to plant this thin and delicate plant in groups.

Growing in containers
Flax grows well in containers.

Healing properties
The edible seeds are rich in vitamins and minerals. Eaten whole or crushed with some water, they are a gentle laxative. With their high mucilage content, they have a soothing effect on the entire digestive system. They are also an effective expectorant and anti-inflammatory. The seeds are high in phytoestrogens, which help minimise menopause symptoms.

Flax seed oil contains high levels of the omega-3 essential fatty acid (alpha-linolenic acid) which has numerous health benefits, particularly in treating cardiovascular problems and lowering cholesterol. The oil is also beneficial for skin conditions such as eczema and psoriasis.

How to use
Fresh: (seeds) For constipation eat 1–2 tablespoons of seeds with 1–2 glasses of water.
Infusion: (ground seeds) For coughs and sore throats; to ease menopause symptoms.
Poultice: (seeds) Crush the seeds to make a poultice for boils, abscesses and stubborn thorns.
Maceration: (seeds) Soak seeds in water until they swell and thicken. Use to treat inflammations of mucous membranes, such as pharyngitis.
Oil: (seeds) Take 1–2 teaspoons daily to lower cholesterol, and to treat eczema, psoriasis, dry itchy skin, rheumatism and arthritis.

Culinary uses
Flax flowers can be added to salads or frozen in ice cubes. Use the golden brown seeds in baking or add to a muesli mix. Sprout the seed for salads.

G

GINGER
Zingiber officinale

Growing
- Tender perennial
- Rich, well-drained soil
- Dappled shade
- 1m high and spreading

Harvesting
In autumn, dig up the whole plant and break up the rhizomes. Place them in the sun for a few days to cure before storing them in a cool dry place.

Parts used
Rhizomes

Uses
- Pungent
- Warming
- Circulatory stimulant
- Prevents nausea
- Promotes sweating
- Antiseptic
- Relieves flatulence
- Relieves spasms
- Increases blood flow to area
- Expels phlegm

Ginger was very popular during Roman times, but with the fall of the Empire it disappeared from European tables. It wasn't until Marco Polo brought it back from his voyages to the east that it began to be used in Europe again.

In the garden
Although ginger is a tropical plant, it is easily grown in home gardens in other areas. It grows about a metre high, with glossy green, strappy leaves and likes warm weather and plenty of moisture. It prefers filtered sun and rich, well-drained soil. Start by buying a few fresh rhizomes in spring from your local greengrocer. Look for pieces with well-developed growth buds. These look like fat, pointed horns sticking out at the end of a piece of ginger. They are paler than the rest of the ginger. Wait until daytime temperatures are over 20 °C before planting the ginger out.

A–Z OF HERBS

A few days before planting, cut the rhizomes into pieces about 5–8 cm long, making sure each piece has at least two growth buds. (Cutting a couple of days before allows the cut surfaces to dry, reducing chances of disease.) Dig a hole about 5–10 cm deep. Place the rhizome at the bottom with the growth buds pointing upwards. (The top should be just below the surface once buried.) Cover with compost and press down firmly. Water well until the ground is soaked around the rhizome. Keep well watered and mulched throughout the growing season and it will spread by growing new rhizomes underground. Towards the end of summer, the leaves will start to die back. In frost-free areas, choose a few fat rhizomes to replant immediately. Keep them well mulched throughout winter and new shoots will pop up in spring.

Growing in containers
Ginger grows happily in pots and can be brought indoors during winter in areas with bad frosts.

Healing properties
Ginger is a warming herb and is particularly good for colds, coughs and chills. It stimulates circulation and makes the body sweat. It contains anti-inflammatory compounds, helping reduce pain and increase mobility in arthritis and other joint problems. It is a cleansing herb, ridding the body of toxins. It also aids digestion and is very good for reducing nausea, especially morning and motion sickness.

How to use
Decoction: (fresh or dried root) For colds and fevers and as an expectorant for congested lungs. To reduce nausea, to relieve indigestion and reduce flatulence. To warm the body and promote sweating. To increase circulation and reduce inflammation.
Tincture: (fresh root) As for decoction, but stronger.

Culinary uses
Although we call it 'ginger root', technically, the part of the ginger plant we eat is a rhizome. This is a swollen underground stem, which has small roots coming off it. Ginger has a wide range of uses in kitchen, from curries and stir-fries to desserts and drinks. It preserves well in syrup and can be eaten raw or cooked.

GOLDENROD
Solidago virgaurea

Growing
- Hardy, herbaceous perennial
- Prefers poorer soils; does not like wet feet
- Full sun
- 1 m high by 60 cm wide

Harvesting
Harvest leaves during summer and flowers in late summer and autumn, as soon as they are fully open.

Parts used
Leaves and flowers

Uses
- Drying
- Astringent
- Relieves inflammation of mucous membranes
- Antifungal
- Antiseptic

This herb brings a bright splash of yellow to the garden in late summer and autumn. According to old European superstition, if goldenrod springs up it indicates there is buried treasure where it is growing.

In the garden
Goldenrod grows easily from rooted cuttings (see page 23) or from seedlings. It spreads by sending out underground runners and, if these are not controlled, it can become invasive. Plant in full sun in well-drained soil. It can grow in quite poor soil. Make sure you remove all the flower heads before they go to seed, otherwise it will seed itself all over your garden. It dies back in winter where there is frost but will pop up again in spring. It is a good companion plant, attracting beneficial insects into the garden, particularly aphid-munching lacewings.

Growing in containers
Goldenrod can grow in a large pot but needs to be repotted every year.

Healing properties
This was a well-known medieval healing herb and was also used by Native Americans. Culpeper believed that goldenrod 'is a sovereign wound herb, inferior to none both for inward and outward hurt'. It is used externally to treat skin diseases, small wounds and insect bites and internally for fevers, hay fever, allergies, catarrh, kidney stones and urinary infections.

How to use
Infusion: (flowers and leaves) Drink 2–3 cups a day to treat urinary infections, kidney stones, stuffy colds, allergies, sinusitus and irritable coughs. Gargle to treat a sore throat. Use as a douche for vaginal thrush.
Tincture: (flowers and leaves) As for infusion, but stronger.
Ointment: (flowers and leaves) As a salve for dry and itchy skin conditions such as eczema and psoriasis; for fungal conditions; and to treat small wounds and insect bites.

Other uses
The yellow flowers of goldenrod can be used to dye natural fibres a rich gold colour.

GOTU KOLA
Centella asiatica

Growing
- Semi-hardy perennial
- Moist, well-drained soil
- Full sun or semi shade
- 40cm high and spreading

Harvesting
Pick leaves as and when needed. The leaves can be dried for later use.

Parts used
Leaves and stems

Uses
- Cooling
- Helps mental functioning
- Rejuvenates brain cells
- Heals wounds
- Relaxes and restores nervous system

Cautions
Excessive use can lead to itchiness and headaches. Avoid this herb if you are pregnant, breast-feeding, have an overactive thyroid or are taking sedatives.

Gotu kola is commonly known as pennywort. Legend has it that the Tai Chi master Li Ching-Yuen was 256 years old when he died in 1933. His secret? Drinking a daily herbal tea that included gotu kola and ginseng.

In the garden
Once established, this spreading, low-growing herb with its kidney-shaped green leaves will need little attention. It is an ideal ground cover in a herb or vegetable garden as it forms a dense mat that retains moisture in the soil. Gotu kola is a semi-hardy perennial, preferring moist, tropical climates. It survives light frost but needs protection against severe cold. Grow it from a seedling or propagate from a runner with at least one nodule on it. Once established, it will spread by sending out runners, but it doesn't become invasive. In summer it bears small clusters of pinky red flowers underneath its leaves.

Growing in containers
Gotu kola will grow in a well-drained medium.

Healing properties
The Chinese name for gotu kola translates into 'the fountain of youth'. A tenth-century king claimed it was gotu kola that gave him the stamina to satisfy all 50 women in his harem. And its Indian nickname is 'food for the brain'. This gives an idea of the properties of this remarkable herb. It increases collagen, speeds healing, reduces scarring and increases hair growth. It detoxifies, and it improves circulation, mental balance and functioning. It is used to aid meditation, relieve stress and increase spiritual understanding.

How to use
Infusion: Drink up to 3 cups a day to relieve stress and anxiety, detoxify the system, strengthen and restore the nervous system, stimulate hair growth, increase mental alertness, encourage balanced moods, fight premature ageing, ease rheumatism and rheumatoid arthritis, improve memory and circulation and treat varicose veins.
Tincture: As for infusion, but stronger.
Compress: Use a strong infusion as a compress to treat varicose veins, psoriasis and eczema. (CONTINUED OVERLEAF)

A Sri Lankan legend attributes the elephant's longevity to a diet of gotu kola

Ointment: To treat sore joints, burns and small wounds, and to heal scars.
Cream: Use as a face and body cream to rejuvenate skin and improve skin tone, and to treat varicose veins, psoriasis and eczema.

Culinary uses

Young gotu kola leaves can be added to salads and older leaves used as a leafy green, quickly stir-fried as an accompaniment to a hot curry. They go particularly well with coconut milk and chilli.

G

GROUND IVY
Glechoma hederacea

Growing
- Hardy perennial
- Most soils
- Full sun to semi shade
- 15 cm high and spreading

Harvesting
Harvest leaves as and when needed. Pick flowers in spring. The leaves and flowers dry well for later use.

Parts used
Leaves and flowers

Uses
- Bitter
- Cooling
- Relieves inflammation
- Digestive
- Detoxifier
- Tonic

A hardy ground cover, with delicate tubular flowers in spring, this sweet-smelling herb attracts butterflies and bees to the garden.

In the garden
Ground ivy is a creeping ground cover that does best in semi-shady, moist areas. It will, however, grow in rocky, poor soil, but then won't spread as much. It forms a scented ground cover, ideal for attracting insects, smothering weeds and retaining moisture in a vegetable or herb garden. In spring it has delicate, violet flowers. It is easily propagated by cutting slips with roots from an existing plant. As with any member of the mint family, it can spread.

Growing in containers
Ground ivy grows happily in containers, especially combined with taller plants.

Healing properties
Ground ivy has been used as a medicinal herb in Europe for thousands of years. Culpeper considered it a cure-all, treating everything from 'inward wounds, ulcerated lungs' and 'gout in the hands, knees or feet' to easing 'all gripeth pains, windy or choleric humours in the stomach'. High in iron and vitamin C, this is a good tonic herb, especially in spring when it has abundant new growth. It treats colds and coughs effectively, soothes the digestive system and helps heal infections, ulcers and wounds. It is also a detoxifier, helping to treat kidney problems.

How to use
Infusion: (leaves and flowers) Internally to treat coughs and colds, as a detoxifying and enriching tonic, to ease indigestion and to treat infections and ulcers. As a gargle for sore throats. As a steam inhalant for sinusitis. As a mouth wash for infections. Externally as a wash for wounds, cuts and sores. A weak infusion as a wash for sore eyes.
Ointment: (leaves and flowers) To treat inflamed skin conditions and wounds, cuts and sores.
Compress: (leaves) To treat wounds, cuts and sores.

Culinary uses
Ground ivy is a bitter-tasting herb, adding a healthy bite to salads. The flowers are very pretty and can be crystallised for cake decorations or frozen in ice cubes.

H

HEARTSEASE
Viola tricolor

Both pansies and violas are derived from the wild viola or heartsease. Multicoloured garden pansies are hybrids originating from crossing wild viola species to produce plants with bigger flowers and various colours. The smaller flowers of the viola are closer to the original wild flower.

Growing
- Hardy annual or short-lived perennial
- Well-drained, moist soil
- Full sun to dappled shade
- 30 cm high by 30 cm wide

Harvesting
Pick leaves when required and harvest roots in autumn. Pick flowers when they are fully opened. All parts can be dried for later use.

Parts used
Flowers, leaves and roots

Uses
- Moistening
- Slightly bitter
- Expels phlegm
- Relieves inflammation
- Diuretic
- Stimulant
- Blood vessel toner and strengthener

Cautions
Avoid high doses of this herb as the saponin content can result in nausea.

In the garden
Pansies and violas are easy to grow from seeds or seedlings. Plant in autumn and you will have bright flowers throughout winter, spring and early summer. These plants prefer a cooler, moist environment and don't mind dappled shade. Dead head the flowers to keep them flowering. If you leave some flowers they will form a beautiful triangular-shaped seedpod. Collect the seeds just as the pod begins to dry and sow them the following autumn.

Growing in containers
Heartsease grows easily in containers in well-drained soil. Don't let them dry out.

Healing properties
Viola tricolor's ancient name, heartsease, could originate from its use in love potions. In Shakespeare's *A Midsummer Night's Dream* this was the plant Oberon squeezed into Titania's eyes, to trick her into falling in love with Bottom. (It must have been potent, as Bottom had the head of a donkey!) Its name could also come from its efficacy in treating heart problems. It does this mostly by strengthening and toning blood vessels and stabilising capillary membranes. It also treats many skin disorders (from nappy rash to varicose veins) and has high levels of saponin, making it an effective expectorant.

How to use
Infusion: Internally to stabilise high blood pressure, for chronic skin problems, to increase circulation and stimulate the immune system.
Tincture: (whole plant) Internally for digestive and lung problems, to strengthen blood vessels.
Syrup: (leaves and flowers) For stubborn coughs.
Wash: (leaves and flowers) For nappy rash and varicose veins.
Cream: (leaves and flowers) For rashes, dry or itchy skin and eczema.
Poultice: (leaves and flowers) Apply to rashes, varicose veins and ulcers.

Culinary uses
Pansy flowers have a mild, grassy flavour and are delicate additions to salads and ice cubes.

Other uses
Heartsease plants are good companion plants in a vegetable garden. As they are quick growing they are good at smothering weeds and go well with slower-growing plants such as onions, carrots and all members of the cabbage family. They also go well with lettuces.

H

HORSERADISH
Armoracia rusticana

For more than three centuries horseradish has been used for everything from back pain to an aphrodisiac. It is a perennial member of the mustard family, with both its roots and leaves being edible.

Growing
- Hardy perennial (leaves die down in frost)
- Well-drained, moist soil
- Full sun to dappled shade
- 60–90 cm high and spreading
- Can be invasive

Harvesting
In early winter, or after the first frosts, dig out the whole plant and harvest the roots. Harvest leaves when young, for eating.

Parts used
Leaves and roots

Uses
- Stimulates circulation
- Antiseptic
- Diuretic

Cautions
Use medicinally with caution. Can blister skin if overused and cause internal inflammation. Avoid medicinal use if thyroid function is low, if pregnant or if you suffer from kidney problems.

In the garden

Horseradish is grown from seedlings or fresh root. Choose a piece of root about 20 cm long and about as thick as a pencil. Bury it at an angle with the narrow end 10 cm deep and the thicker end 5 cm below the surface. Keep it well watered until shoots appear. Leave it for the first season to spend the summer building up its root system. Once big enough, it will spread its roots underground and new shoots will start sprouting around the base of the plant. When this happens, begin harvesting by digging these roots out. Horseradish will spread quite quickly unless you keep harvesting the side roots. Don't worry about killing it off – this plant can take quite a bit of abuse!

After harvesting, leave a few roots in the ground for next spring. If you don't harvest the whole plant, it will die down during winter but will send up new shoots in spring. If you are worried about horseradish spreading too much, plant it in a bottomless pot sunk into the ground. This will allow its long edible roots to develop but prevents it spreading.

Growing in containers

Horseradish is suitable only for a large, deep container.

Healing properties

Pungent horseradish root was used medicinally long before it became a condiment for meat and fish. It stimulates circulation and is a digestive, antiseptic and diuretic. It contains mustard oil, which helps clear sinuses, ease chest congestion and asthma.

How to use

Infusion: (roots) Internally to treat asthma, congested chest and colds and flu, clear water retention, increase circulation, as a digestive stimulant and to cleanse the liver.
Poultice: (grated roots) As a poultice to treat rheumatism and muscle pain and aches.

Culinary uses

The roots are used to make hot horseradish sauce and a few young leaves add a tangy bite to salads and stir-fries. Freshly harvested roots make a far stronger horseradish sauce than most commercial varieties. The roots should be processed as soon as possible after harvesting. Scrub them clean and blend them in a food processor – but be careful, it is a pungent process. Add 2–3 tablespoons of vinegar per 1 cup of horseradish. If you want a milder sauce, then add it immediately. If you want it hotter, then wait about three minutes. It will keep, sealed in jars, for up to six weeks or in the freezer for up to three months.

Other uses

Scientists have conducted successful experiments using horseradish to decontaminate wastewater. Horseradish also has potential as a cleansing green manure for contaminated soil. Horseradish spray is effective against many leaf-eating bugs and fungal diseases.

HOUSELEEK
Sempervivum tectorum

Growing
- Hardy perennial
- Dry, well-drained soil
- Full sun
- 5–8 cm high and spreading

Harvesting
Pick leaves when needed.

Parts used
Leaves

Uses
- Cooling
- Astringent

Cautions
Houseleek can cause diarrhoea if too much is taken internally.

Also known as hen and chicks, this is an ancient magical herb, used for protection against both witchcraft and lightning. For centuries this fleshy plant was grown on roofs to protect houses against fire and it is interesting that the houseleek is the most popular plant used today for growing a green roof.

In the garden
This is a ridiculously easy herb to grow and maintain. All it needs is a tiny bit of ground to cling to and a sniff of water occasionally and it will survive tenaciously. Grow it by simply breaking off a small offshoot and planting it in full sun, in well-drained soil.

Growing in containers
Houseleeks like containers filled with a very well-drained, not-too-rich growing medium. Do not over-water.

Healing properties
Similar to aloe, the gel inside the leaves is very soothing for insect stings, burns, rashes and bites. The gel also softens warts. Internally it can be taken to treat a mouth or throat infection and to ease chest problems.

How to use
Fresh gel: Break open a leaf and apply directly to burns, stings, rashes and bites. Apply regularly to warts to soften.
Infusion: Taken internally to ease bronchitis. Use as a gargle to treat mouth and throat infections.

L

LAVENDER
Lavandula angustifolia

Growing
- Hardy perennial
- Well-drained soil
- Full sun
- Sizes vary with variety

Harvesting
Cut flowers just as they open. Pick leaves when required.

Parts used
Flowers (medicinal and culinary) and leaves (culinary)

Uses
- Bitter
- Drying
- Cooling
- Relaxant
- Relieves spasms
- Nervous system tonic
- Circulation stimulator
- Antibacterial
- Pain reliever
- Antiseptic
- Relieves inflammation

Cautions
Lavender is a uterine stimulant so avoid high doses during pregnancy.

Lavender is a popular and ancient herb. There are many different varieties used for ornamental, culinary and medicinal purposes. Its name comes from the Latin word *lavare* meaning 'to wash'. It also has the reputation of bringing good luck in affairs of the heart. To win the love of the one you desire, you should carry a sachet of lavender flowers close to your bosom.

In the garden
A native of the Mediterranean, this hardy perennial likes full sun and is drought tolerant. However, it adapts well to a range of climates. It flowers nearly all year round, attracting many beneficial insects with its fragrant purple spears. Its strong-smelling leaves repel aphid, whitefly and other harmful insects. It is also a rodent repellent and if rats are a problem, a lavender hedge around a vegetable garden is a good deterrent.

Grow lavender from seed in seed trays or purchase seedlings. It dislikes being damp and if its roots are constantly wet it could die. Plant it in well-drained soil with plenty of space for air circulation. Prune in early spring, cutting back about 8 cm of growth. Cut the flowers regularly and it will keep producing more.

Growing in containers
Lavender will grow happily in a large container with well-drained growing medium.

Healing properties
Throughout history lavender has been used as a calming, soothing herb. It has many beneficial properties: antiseptic, anti-inflammatory, antibacterial, antifungal, anticonvulsive and antidepressant. It also treats burns, insect bites and small wounds. Only the flowers are used medicinally.

How to use
Infusion: For headaches, depression and nervous tension, hyperactivity, insomnia, neuralgia, indigestion and colic (especially due to tension).
Tincture: For headaches, hyperactivity and depression.
Mouthwash: To cleanse the mouth and freshen breath.
Cream: Either an infusion or a few drops of essential oil in a cream as a burn and sunburn soother.
Ointment: Add a few drops of essential oil to treat eczema and fungal conditions.

(CONTINUED OVERLEAF)

Tuck a sprig of lavender under your pillow before you go to sleep and make a wish; your dreams will reveal if it will come true

Essential oil: Use in a massage oil for sore muscles and to ease headaches. Add to a bath or oil burner to ease nervous tension. Add to a chest rub to ease bronchial spasms and asthma. Inhale steam to ease an asthma attack.

Infused oil: As for essential oil.

Culinary uses

Lavender is a strongly flavoured herb so use it sparingly in the kitchen. The flowers and leaves are delicious in baking, jams, jellies and cordials.

Other uses

Lavender is a useful insect repellent in the garden and home. Add it to a moth-repellent mix (see page 213 for a recipe) or rub fresh flowers onto your skin to keep flies away. Toss lavender leaves in the hens' run to keep their bedding fresh.

LEMON BALM
Melissa officinalis

Growing
- Hardy perennial
- Moist soil
- Full sun to partial shade
- 75 cm high by 45 cm wide and spreading

Harvesting
Pick leaves when needed and flowers as they open. The leaves do not dry well and are best used fresh. To preserve leaves for use in winter, rather freeze them.

Parts used
Leaves and flowers

Uses
- Cooling
- Drying
- Slightly bitter
- Sedative
- Lowers fever
- Antidepressant
- Antiviral
- Relieves spasms
- Antibacterial
- Relieves flatulence

Lemon balm, also known as melissa or bee balm, bears small white flowers that bees love. Ancient Greek beekeepers would crush the leaves of this herb and rub them on the hives to encourage bees to return. (*Melissa* means 'honeybee' in Greek.)

In the garden
This lemon-flavoured member of the mint family is a worthy addition to the herb garden. It is a hardy plant requiring nothing more than the occasional trimming. It likes full sun to partial shade and plenty of moisture. After it has flowered, cut it back to prevent it becoming straggly. Lemon balm can be invasive, so keep an eye on it and divide the plant if necessary (see page 25). Its strongly scented leaves repel many pests. Sprinkle the leaves around the base of vulnerable seedlings or on newly seeded areas. It is a good companion for tomatoes and squash plants.

Growing in containers
Lemon balm does well in containers and this prevents it spreading.

Healing properties
Lemon balm's usage to reduce fevers and as a calmative and antidepressant dates back hundreds of years. It lowers blood pressure and treats colds and flu. It is good for any problems related to nervous tension such as insomnia, tension headaches, stomach ailments and depression.

How to use
Infusion: For depression and any disorders related to nervous tension. To prevent insomnia and hyperactivity. To reduce blood pressure. Drink at the first sign of cold or flu to help prevent its onset.
Tincture: As for infusion, but stronger.
Cream: Insect repellent.
Ointment: (use an infusion or essential oil) For cold sores, minor wounds and insect bites.
Compress: To relieve painful joints and gout.
Infused oil: Use as a massage to relieve depression and tension or as a chest rub to relieve bronchitis and asthma. (CONTINUED OVERLEAF)

Wine infused with lemon balm mends a broken heart and clears the way for new love to enter

Culinary uses

Lemon balm doesn't take well to cooking and is best eaten fresh. It adds a lemony flavour to salads and soups, and goes well with fish dishes. It can be used to make a lemon-flavoured vinegar. It also goes well with cheese dishes and is a great addition to a cream cheese dip.

Other uses

Lemon balm is a useful insect repellent in the garden and home. Add it to a moth-repellent mix (see page 213 for a recipe) or rub fresh flowers onto your skin to keep insects away.

L

LEMON GRASS
Cymbopogon citratus

Growing
- Perennial; or annual in very cold areas
- Well-drained soil
- Full sun
- 1.7m high by 1m wide

Harvesting
Harvest lemon grass by cutting stalks at the base. Choose the thicker, older stalks first.

Parts used
Stem and leaves

Uses
- Sedative
- Calming
- Painkiller
- Reduces fever
- Astringent
- Antidepressant
- Antifungal
- Antibacterial
- Cleansing

Cautions
Do not use medicinally while pregnant or breast-feeding. Do not use around the eyes. It may cause irritation on sensitive skin.

Lemon grass is an essential ingredient in many Southeast Asian dishes with its unmistakable fresh, lemony taste. Native to Sri Lanka, it is a fast-growing perennial. It has long, strap-like green leaves and can grow over a metre tall.

In the garden
Although lemon grass is native to tropical areas, it grows easily in a wide range of climates. It is a clump-forming grass with cane-like stems and long strappy leaves. Plant lemon grass towards the back of the herb garden where it won't block sun from reaching shorter plants. Even though its growth slows down during colder weather, it will survive mild frosts. In spring, cut back dead leaves and it will start growing again. It is a good idea to divide it every couple of years (see page 25). It's a tough plant, not bothered by any insects or pests (although dogs do like chewing it). It is a good companion for many plants, particularly brassicas, as its strong scent repels harmful insects.

Growing in containers
Lemon grass grows well in containers.

Healing properties
Lemon grass has calming, sedative and antidepressant properties and reduces fevers and soothes stomach cramps. The essential oil is an antiseptic and is effective against fungal and bacterial infections.

How to use
Infusion: To calm any stress-related disorders and soothe the digestive system.
Tincture: As for infusion, but stronger. Reduces fever.
Ointment: To relieve backache and painful joints, rheumatism, muscular pains and sprains.
Essential oil: Add to an ointment to treat fungal and bacterial infections.
Infused oil: Use as a massage oil to relieve stress, treat painful backache and joints and to ease muscular pains and sprains.

Culinary uses
Use whole lemon grass stems in curries and soups (bruise the stem first, by hitting it with the handle of a chef's knife). Or peel away the hard outer layers and finely slice the tender inner stem for use in stir-fries and curries. (CONTINUED OVERLEAF)

If you feel creatively blocked, burn some lemon grass to help stimulate the left side of your brain

It is also a delicious addition to desserts – lemon grass sorbet is particularly refreshing. The leaves are full of flavour. Add them to a jug of drinking water or to vegetable-steaming water.

Other uses
Lemon grass is a good insect repellent. Use essential oil in a burner to keep insects away. This also clears the smell of cigarette smoke or animals from a room.

LEMON VERBENA
Aloysia triphylla

Growing
- Semi-hardy perennial
- Well-drained soil
- Full sun
- 2 m high by 1.5 m wide

Harvesting
Pick leaves throughout summer. The flavour keeps well when dried.

Parts used
Leaves

Uses
- Sedative
- Calming
- Relieves flatulence
- Decongestant
- Aids digestion
- Antibacterial
- Antiseptic
- Relieves spasms
- Reduces fever

This fast-growing, aromatic herb is a native of South America. It is seldom found at the greengrocer, so it is worth growing your own. A cleansing herb, it is used as incense to break up old patterns and habits, and to clear away unwanted things.

In the garden
Lemon verbena likes full sun and will easily grow up to two metres tall in one season. Plant it at the back of your herb garden where it won't block light. If you want a bushier shrub, cut off the top of the main stem to encourage side sprouting. It likes well-drained soil and is quite drought hardy. If it dies down during winter, give it a severe pruning in early spring after the last danger of frost has passed. It grows easily from slips. This is a tropical plant that will not survive severe frosts.

Growing in containers
Lemon verbena grows well in a pot, but keep it trimmed. It must be well mulched in winter.

Healing properties
Lemon verbena has a calming effect, particularly on the digestive system, reducing spasms and colic. It also reduces fever.

How to use
Infusion: To treat feverish colds and ease asthma. To calm any nervous disorders, particularly in the digestive system. Reduces indigestion, colic, flatulence and diarrhoea (especially related to a nervous condition).
Tincture: As for infusion, but stronger.
Ointment: To treat boils and acne.

Culinary uses
Lemon verbena has a fresh, sweet, lemon fragrance. Add fresh leaves to steaming water to give vegetables a lemony touch or add to a jug of drinking water. Use to create flavoured sugar and vinegar. Add chopped leaves to soups, rubs, fruit salads and fish dishes or steep leaves for a refreshing digestive tea.

Other uses
Lemon verbena keeps clothes in storage fresh and helps to deter moths as it is an effective insect repellent.

L

LOVAGE
Levisticum officinale

This bushy herb looks and tastes very similar to celery – but is much easier to grow. As its name indicates, it is used in love potions and charms, and is said to attract new love or increase passion in an existing relationship.

Growing
- Hardy perennial
- Moist, fertile soil
- Full sun or light shade
- 1.5 m high by 1 m wide and spreading

Harvesting
Harvest leaves in spring to early summer, before flowering. (After the plant flowers, the leaves taste bitter.) It does not keep its flavour well when dried, so rather freeze it for winter use. Harvest seeds after flowering. Harvest the root from two- or three-year-old plants in autumn.

Parts used
Leaves, seeds and roots

Uses
- Warming
- Relieves flatulence
- Aids digestion
- Relieves inflammation
- Antioxidant
- Reduces water retention
- Relieves spasms
- Digestive
- Diuretic
- Expels phlegm
- Stimulant

In the garden
Lovage is a close relation to both angelica and celery. Like their relatives, this herb prefers a rich, moist growing environment in sun or light shade. It will die down completely in winter and benefits from being well mulched during its dormancy. Large plants can be divided in spring. The yellowish-white flowers attract plenty of beneficial insects to the garden.

Growing in containers
Lovage grows well in a container, but keep it trimmed.

Healing properties
Lovage has been used for hundreds of years to treat digestive problems, ease colic and reduce flatulence. It is a good blood cleanser and can be used to treat liver problems and for skin eruptions, gout and rheumatism. Its cleansing properties are also good to treat skin problems such as acne. It contains high levels of quercetin (an anti-inflammatory and antioxidant), making it an effective herb for treating allergies and asthma.

How to use
Infusion: (leaves and seeds) As a digestive (especially after a heavy meal), to encourage menstruation, to ease asthma and allergies, to cleanse liver and blood, to treat colic and flatulence and as a mild expectorant.
Decoction: (roots) To treat sore throats. Externally to treat skin problems such as boils, acne and pimples.

Culinary uses
Soups, stocks and stews all benefit from the addition of lovage. Young leaves can be added to salads and hollow stems used as drinking straws for a Bloody Mary. The stems can be steamed as a vegetable and the seeds can be dried and added to rubs, used in baking or ground up for an interesting flavour addition to soups and stews.

Cautions
Do not use if pregnant or if you have kidney problems. Long-term use can increase sensitivity to the sun.

MARJORAM AND OREGANO
Marjorana hortensis and *Origanum vulgare*

Growing
- Annual and perennial
- Well-drained soil
- Full sun
- Height and width depends on varieties

Harvesting
Pick leaves in spring and early summer, before the plants begin to flower. The leaves retain flavour well when dried. Pick flowers before they go to seed.

Parts used
Leaves and flowers

Uses
- Relaxing
- Warming
- Calming
- Antiviral
- Antibacterial
- Relieves inflammation
- Relieves spasms
- Relieves flatulence
- Antioxidant
- Antifungal
- Expels phlegm
- Antiparasitic

Legend has it that Aphrodite, the Greek Goddess of Love, created these fragrant herbs. Whatever their origin, no herb garden should be without the easy-to-grow and endlessly useful marjoram and oregano.

In the garden
Both these herbs have a beneficial effect on any neighbouring plant. Marjoram is a species of oregano and both are in the mint family. Marjoram is more sensitive to frost and is mostly grown as an annual. Oregano is hardy.

All varieties, annual or perennial, from golden creeping groundcover to upright bushy plants, like to grow in hot sunny places and don't require much water. Although they can be grown from seed, it is much easier to grow them from seedlings. Once you have a healthy plant, rooted cuttings can easily be transplanted to other parts of the garden (see page 23). These are very undemanding herbs to grow. All they need is to be trimmed back every now and then so they don't become straggly or too bushy.

Growing in containers
Marjoram and oregano both grow easily in containers. They need a well-drained growing medium and don't like being over-watered.

Healing properties

Both herbs have a high thymol content, making them effective antiseptics. Marjoram's healing properties include treating asthma, hay fever, sinus congestion, colds, coughs, indigestion, stomach pain, headaches and nervous conditions. Oregano's healing properties include treating indigestion, bloating and flatulence, coughs and bronchial problems, urinary problems and headaches, and promoting and regulating menstruation. It is a natural antihistamine, effective in treating allergies. Oregano also has expectorant and decongestant properties, making it a useful herb for congested respiratory problems. Both herbs have pain-relieving qualities, and are used externally to treat painful muscles, rheumatism and swollen, sore joints. The essential oil of both herbs effectively relieves toothache.

How to use marjoram

Infusion: (leaves and flowers) To treat asthma, hay fever, sinus congestion, colds, coughs, indigestion, stomach pain, headaches and nervous conditions.
Mouthwash or gargle: (leaves and flowers) To keep mouth clean and as a preventative measure against coughs and colds.
Essential oil: To treat toothache; in an oil burner or added to bath oil to calm nervous conditions.
Infused oil: (leaves and flowers) As for essential oil and cream.

How to use oregano

Infusion: (leaves and flowers) To treat the digestive system; to alleviate hay fever and allergies, coughs, bronchial problems and congested sinus; to relieve headache pain; and to promote and regulate menstruation.
Mouthwash or gargle: (leaves and flowers) To keep mouth clean and as a preventative measure against coughs and colds.
Cream: (leaves and flowers) To treat itchy skin, fungal conditions, minor wounds and sores.
Poultice: (leaves) To treat sore muscles, swollen joints and pain from rheumatism.
Essential oil: To treat toothache; in an oil burner or added to bath oil to calm nervous conditions.
Infused oil: (leaves and flowers) As for essential oil and cream.

Culinary uses

Marjoram and oregano originate in the Mediterranean and they are both used extensively in Mediterranean cooking. Marjoram is not as robustly flavoured as oregano. It is sweeter, lighter and does not keep its flavour well when cooked. It is better added at the end of cooking or used uncooked. Oregano is stronger than marjoram and can withstand longer cooking. The hotter and drier the weather, the heartier its flavour.

Cautions
Do not use the essential oils during pregnancy.

Like fraternal twins, these two herbs are very similar — but with subtle differences: marjoram (above) is milder and more complex while oregano (opposite) is sharper and zestier

M

MILK THISTLE
Silybum marianum

This is an unusual and surprisingly beautiful herb with its green and white mottled leaves and purple flowers. It is also known as Our Lady's thistle and St Mary's thistle. The story goes that when a drop of the Virgin Mary's milk fell on a nearby thistle, it imbued it with healing properties and created the characteristic 'spilt milk' pattern.

Growing
- Hardy annual or biennial
- Any soil
- Light shade to full sun
- 1.5 – 3 m high by 50 cm – 1.5 m wide

Harvesting
Harvest the leaves when they are young. Harvest flowers in full bloom and seeds once the seed ball has formed. Beware, the whole plant is covered in prickles and the seed ball is particularly thorny. It is a good idea to arm yourself with leather gloves and long sleeves before harvesting this prickly plant. Cut the seed ball open and extract the seeds with tweezers.

Parts used
Leaves, flowers, roots and seeds. (The seeds are the most effective medicinally.)

Uses
- Bitter
- Diuretic
- Liver tonic
- Stimulates bile
- Relieves inflammation
- Antioxidant
- Mild laxative
- Increases lactation

In the garden
If you have seeds saved from an existing plant, sow them in early summer at a depth of about 3 mm. Seedlings are available from specialist herb nurseries. Milk thistle is not fussy about its soil and will grow in light shade to full sun. It contains high levels of potassium and its leaves are a good addition to the compost heap.

Growing in containers
Milk thistle is suitable for growing in a large pot with a well-drained growing medium.

Healing properties
Milk thistle has been used for thousands of years to treat the gall bladder, liver and kidneys. It encourages liver cell renewal and repair and is particularly effective in treating liver congestion and cirrhosis. It is a powerful antioxidant and toxin cleanser and is helpful in reducing the side effects of chemotherapy. It is an effective anti-inflammatory. Silymarin, the most active ingredient in milk thistle, is not highly water-soluble; therefore a tincture made with alcohol is stronger. An interesting use for milk thistle is to treat mushroom poisoning – a useful thing to know if you are planning some foraging!

How to use
Infusion: (leaves and seeds) To increase lactation, as an antioxidant and cleansing tonic; to treat digestive problems and as a mild laxative.
Tincture: (leaves, ground seeds and flowers) To treat liver and gall bladder diseases, digestive problems and poisoning; to reduce effects of chemotherapy and to increase lactation.

Culinary uses
Young milk thistle leaves can be eaten as a green – make sure you remove the spines first! Tender roots of year-old plants roots can be steamed and the flower buds are similar to artichokes.

MINT
Mentha species

Mint is an easy herb to grow — one of the most difficult things about it is preventing it from taking over your garden. This is a protective herb, with the reputation of warding off malignant energy and intentions.

Growing
- Hardy to semi-hardy perennial
- Prefers moister soil
- Full sun to semi shade
- Various heights and spreading

Harvesting
Harvest fresh leaves as required. In late spring to early summer, when mint is at its tastiest and most abundant, harvest leaves and dry them to be bottled for use in midwinter.

Parts used
Leaves and flowers

Uses
- Pungent
- Cooling
- Drying
- Relieves spasms
- Antiseptic
- Digestive
- Relieves flatulence
- Prevents vomiting
- Promotes bile
- Pain reliever

In the garden
Mint likes full sun and plenty of water. Some varieties, such as apple mint, don't mind semi shade. In cold winter areas, cut it right back in winter and mulch its roots well. In spring it will bounce back vigorously. There are many varieties of mint, with flavours ranging from pineapple to chocolate. Mint spreads by underground runners and can be very invasive. Either plant in pots above ground or, if you want it as part of your herb garden, then bury bottomless pots in the ground. Check the pots every year as mint runners are strong and can break through a pot. Mint grows well with tomatoes and members of the brassica family. Don't plant with cucumber.

Growing in containers
Mint grows well in containers and this is a good way to prevent it taking over.

Healing properties
There is a good reason why after-dinner mints are offered after dinner. One of mint's main benefits is as a soothing digestive. Peppermint (*Mentha x piperita*) is the variety widely used medicinally. It is an antispasmodic, relaxing the colon muscles and digestive tract, and stimulates bile. It effectively treats indigestion, stomach cramps, flatulence, heartburn and nausea. It has a beneficial effect on irritable bowel syndrome, constipation and abdominal pain. It also promotes sweating and can be used to treat fevers. Externally mint can be used to ease joint and muscle aches. Added to a steam inhalation it eases nausea and decongests nasal passages.

How to use
Infusion: (leaves and flowers) To treat indigestion, insomnia (especially if caused by indigestion or stomach tension), constipation, irritable bowel syndrome, colic, flatulence, nausea, travel sickness and feverish conditions. As a wash for itchy skin or as a mouthwash to freshen the breath.
Tincture: (leaves and flowers) As for infusion, but stronger.
Steam inhalation: (leaves and flowers or 1–2 drops of essential oil) To ease nausea and decongest sinuses.
Ointment: (leaves and flowers) Add a few drops of essential oil to an ointment to treat skin irritations, neuralgia, burns and inflammation.
Compress: (leaves and flowers) Soak a pad in a hot infusion and apply to painful joints and muscles or to the temples to ease a headache.
Essential oil: As for infused oil but stronger.
Infused oil: (leaves and flowers) Rub on painful joints and muscles; rub onto temples to ease headaches; rub onto irritated skin, burns, neuralgia and inflammation.

Culinary uses
Mint is a fresh and versatile herb in the kitchen. It does not retain its flavour well if cooked too long and is best added at the end of cooking. It pairs well with strong, meaty dishes such as roast lamb or kofte and adds a bright note to salad dressings and desserts. Chop it into a yoghurt raita or tomato relish to balance a hot curry or mix it with salty cheeses such as feta and haloumi. It goes well with various cocktails and fruit drinks. It is also a classic paired with chocolate or made into jelly.

Other uses
Mint deters cockroaches and ants. Mix essential oil with some water and wipe or spray it on surfaces where they are active. Add essential oil to a body cream to create an insect repellent. Add to an oil burner to keep flies and mosquitoes away.

Cautions
Do not over-use as a steam inhalant as it can irritate mucous membranes. Do not take medicinally while breast-feeding as it can reduce milk flow.

MOTHERWORT
Leonurus cardiaca

Growing
- Hardy perennial
- Well-drained soil
- Full sun to semi shade
- 1 m high and spreading

Harvesting
Pick leaves as and when needed. They dry well for later use. Pick flowers in midsummer for drying.

Parts used
Leaves and flowers

Uses
- Regulates menstruation
- Soothes the nerves
- Uterine stimulant
- Heart strengthener and regulator

Cautions
Do not take when pregnant as it is a uterine stimulant. The leaves can cause a rash.

Both its Latin and common name indicate this plant's ancient usage as a treatment during childbirth and as an effective heart-strengthening remedy. Like other members of the mint family, motherwort is a protective herb, particularly effective for mothers and children.

In the garden
A member of the mint family, motherwort can spread quickly. In colder areas it will die back after frosts but will reappear as soon as spring warms up. It is an attractive plant with pretty pink flowers in summer, which attract beneficial insects. Grow this hardy perennial from seedlings or divide from an existing plant (see page 25) and position in full sun to semi shade in well-drained soil.

Growing in containers
Motherwort grows well in containers. Do not over-water.

Healing properties
Motherwort has been used to treat many female problems such as menstrual cramps, childbirth, post-partum depression and menopause. It is a calming herb that soothes nervous conditions such as insomnia, heart palpitations and rapid or irregular heart rate. It also brings down high blood pressure.

How to use
Infusion: To regulate menstruation, strengthen the uterus and uplift spirits after childbirth, ease menstrual cramps, calm nervous conditions and insomnia, regulate and strengthen the heart and reduce blood pressure.

Other uses
Similar to mugwort, motherwort is used to promote clear dreaming.

MUGWORT
Artemisia vulgaris

Growing
- Hardy perennial
- Well-drained soil
- Full sun to semi shade
- 30 cm high and spreading

Harvesting
Pick leaves as and when needed. They dry well for later use. Harvest the root in autumn for drying.

Parts used
Leaves and roots

Uses
- Warming
- Digestive
- Soothes the nerves
- Reduces fevers
- Regulates menstruation

Cautions
Do not take when pregnant.

Mugwort is an ancient magical herb, used to fend off evil spirits and venomous thoughts. Travellers also put it in their shoes to prevent them becoming weary on long journeys.

In the garden
Mugwort can become a bit of a menace in the garden as it spreads quickly. Either grow it in a container or bury it in a bottomless pot to prevent it spreading. It doesn't mind poor soil as long as it is well drained. Plant in full sun or semi shade. It will die back in winter in colder areas, but pops up again in spring. It is a straggly perennial that benefits from being trimmed.

Growing in containers
Mugwort grows happily in containers. Do not over-water.

Healing properties
Similar to other members of the artemisia family, slightly bitter mugwort stimulates appetite. It is also a calming herb for nervous conditions. It regulates menstrual flow and sweating.

How to use
Infusion: (leaves) To regulate menstruation (especially for young women just starting); to aid digestion; to reduce fevers (especially at the onset of an illness); to calm jittery nervous conditions and insomnia.
Decoction: (roots) As for infusion, but stronger.

Other uses
Mugwort is a powerful metaphysical herb. It is a common ingredient in smudge sticks, used to smoke out evil spirits and stagnant energy in houses. It is best known as a dream herb, enabling you to remember your dreams more clearly as well as enhancing dreams and trances. For this purpose it can be taken internally, or fresh leaves can be placed under the pillow.

M

MUSTARD
Brassica var. *juncea, alba, nigra*

Mustard has been used for thousands of years for its pungent flavour and medicinal properties.

Growing
- Hardy annual
- Well-drained soil
- Full sun to semi shade
- 80 cm – 1 m high by 30 cm – 1 m wide

Harvesting
Pick leaves as and when needed. Harvest the seedpods when they begin to turn yellowy-brown. If you leave them too long they will burst.

Parts used
Leaves, flowers and seeds

Uses
- Warming
- Antiseptic
- Antibacterial
- Antifungal
- Aids digestion
- Increases circulation

Cautions
The same effect that causes the warming can also burn the skin. Excessive use can cause irritation, inflammation and blisters. Use with caution, especially on sensitive skin.

In the garden

Mustard is an invaluable multiuse crop for small gardens, as the leaves, flowers and seeds are all edible. It is quick growing and can be used as a cover crop to smother weeds and as a green manure. Mustard is very easy to grow – almost too easy as it can become a bit of a self-seeding weed if not controlled. It is a hardy annual that grows throughout the year.

There are different varieties of mustard but they all have similar growing requirements. Sow the seed directly. The plants like full sun or light shade in full summer and sufficient moisture. If using mustard as a green manure, dig it in before it sets seed.

Growing in containers

Mustard grows happily in containers. Move them into a spot with afternoon shade during summer.

Healing properties

Mustard seed has been used as a healing herb since ancient times. It contains high levels of sulphur and other compounds that have strong antifungal and antibacterial properties. When the seeds are ground and mixed with a liquid, pungent enzymes are released that stimulate circulation and healing. These warming effects successfully treat rheumatism, arthritis, chilblains, joint pains, sore feet, backache, sore muscles and chest colds and coughs. Mustard also eases constipation and aids digestion.

How to use

Infusion: (crushed seeds) To speed up the metabolism; to stimulate circulation and appetite; to ease coughs, colds and constipation; to lessen asthma symptoms.
Foot bath: (1 teaspoon mustard powder to 2 ½ cups boiling water) To stimulate circulation, prevent colds and to treat chilblains.
Poultice: (crushed seeds) Also known as a mustard plaster. Make a thick paste with crushed mustard seeds and egg white. Place between two layers of clean cloth. Apply to the affected area, checking often that the skin does not react. To treat rheumatism, arthritis, chest colds, backache and joint pains.
Infused oil: (crushed seeds) Add to a cream to treat fungal skin conditions.

Culinary uses

The Romans were the first to grind mustard seeds to make the hot condiment so familiar today. When we grow mustard in our herb gardens, that biting flavour can be used in a multitude of ways. Add young leaves to salads and steam or stir fry larger leaves. The flowers add a mild mustard flavour to salads and savoury dishes. Mustard seed can be used in pickles, chutneys and curries. To make fresh mustard, grind dried seeds into a powder and add water.

N

NASTURTIUM
Tropaeolum majus

Growing
- Tender annual
- Doesn't mind poor soil
- Full sun to semi shade
- Spreading and rambling

Harvesting
For culinary purposes, pick flowers and leaves throughout the growing season. Harvest seeds while either green or ripe. For medicinal purposes, the leaves should be picked before flowers form.

Parts used
Flowers, leaves and young seeds

Uses
- High in vitamin C and iron
- Antiseptic
- Digestive

With its brilliant sunset colours adding vibrancy to the garden, this is one of my favourite multipurpose plants. Native to Peru, nasturtiums were brought to Spain in the late fifteenth century by the conquistadors. They quickly became popular in the royal European gardens. From the original deep yellow, semi-trailing vine, plant breeders have developed a wide range of colours and varieties.

In the garden
This is one of the easiest plants in the world to grow – just push a seed into the ground during spring and summer and it will sprout. Nasturtiums are either upright or trailing and the latter variety is particularly good in combination with beans and gemsquash on tripods or on an archway. You need to encourage them to grow upwards but they take to it very well. In richer soil they produce fewer flowers and more leaves. In colder climates they die back in the frost but seeds from the previous summer will sprout as soon as the weather warms up.

Nasturtiums make other plants grow more vigorously and prolifically. This is particularly true for tomato, cabbage, cucumber, squash and broccoli. They repel some insects (such as whitefly) and act as a trap crop for others (such as aphids). With their saucer-shaped leaves they are an excellent camouflage plant, confusing harmful insects by disguising the shapes of their target plant.

Growing in containers
Nasturtiums grow happily in containers. Trailing varieties grow well on the edges of hanging containers.

Healing properties
Known as 'Mother Nature's antibiotic' because of the high levels of vitamin C in its leaves, nasturtium is a very useful plant for fighting off colds and assisting the immune system. The leaves contain high levels of iron and have antiseptic and digestive properties. They are most effective when used fresh.

How to use
Infusion: (flowers, leaves and seeds) To stimulate the digestive system and to treat colds and respiratory problems.
Poultice: (seeds) Crushed and used to treat boils and sores.

Culinary uses
The leaves and fresh seeds of nasturtiums have a bright, peppery flavour and the flowers are crunchy and taste a bit like watercress. They can be added to salads, or used as an edible garnish on many dishes. Try stuffing flowers with cream cheese mixed with garlic chives and serving them as part of a cheese platter. The leaves, torn into pieces, add a peppery tang to salads. The seeds, picked when green, can be chopped up and mixed into herb butter. The whole pods can also be pickled for a home-grown version of capers.

NETTLE
Urtica dioica

This valuable plant has been used since the Bronze Age for food, fabric and medication. A rich source of chlorophyll, it was also used as a green clothing dye. In the Second World War, British uniforms were dyed a dark camouflage green with nettle juice.

Growing
- Hardy perennial
- Any soil
- Semi shade
- 1.2 m high and spreading

Harvesting
It is not called stinging nettle for nothing – wear gloves and long sleeves. Nettles cause blisters that really hurt. Harvest young leaves in spring for eating, no more than 10 cm long. To harvest flowers and larger leaves for medicinal use, pick as flowering starts. Dig up roots in early summer before the plant flowers. Dry roots, leaves and flowers for use throughout the year.

Parts used
Flowers, leaves and roots

Uses
- Cooling
- Drying
- Astringent
- Diuretic
- Tonic
- Stimulates circulation
- Stops bleeding
- Promotes lactation
- Lowers blood sugar levels

In the garden
Despite being a stinging plant, nettle is a beneficial addition to the garden. It attracts pollinating butterflies and is a nutrient accumulator, enriching and energising soil. In the compost pile, it activates and speeds up decomposition. Nettles grow easily in any soil and prefer semi shade. Seedlings are available from herb centres or they can be propagated from a piece of root. Nettles spread easily and can become invasive if not cut back regularly. In late autumn, cut the plants back right down to the ground and mulch lightly.

Growing in containers
Nettle grows well in containers. This is a good way to prevent it becoming invasive.

Healing properties
Nettle is rich in vitamins and minerals, making it a good tonic and nutritional herb. It is a traditional blood-cleansing remedy and is a gentle diuretic, ridding the body of toxins. Nettle contains proteins that balance the immune system and also inhibit a wide range of viruses. Nettle effectively treats allergies and rashes, arthritis, gout and rheumatism. It contains high levels of iron as well as vitamin C, making it an effective treatment for anaemia, as the vitamin C ensures the iron is absorbed. Nettle is an excellent hair tonic and growth stimulant and helps keep bones strong as it contains high levels of calcium, magnesium and phosphorous. Nettle also contains high levels of silica, making it good for teeth, nails and hair.

Cautions
Watch out for the stings!

How to use
Infusion: (leaves and flowers) To stimulate circulation and increase lactation; to cleanse the system and treat viral infections; to ease hay fever; to regulate blood sugar; to treat arthritis, osteoporosis, gout, rheumatism and eczema. To strengthen teeth, nails and hair.
Decoction: (root) As a hair rinse to treat thinning hair and dandruff. To strengthen teeth, nails and hair.
Tincture: (leaves and flowers) To treat arthritis and skin problems. To strengthen teeth, nails and hair.
Juice: (young leaves only) As a tonic and to treat anaemia.
Wash: (leaves and flowers) Apply to burns, bites, wounds and eczema. It also soothes the pain caused by stinging nettles!
Compress: (leaves and flowers) Apply to sore joints and sprains, gout, sciatica and tendonitis.

Culinary uses
Nettle is a common weed throughout much of the temperate regions of the northern hemisphere, and young nettle leaves are eaten in spring as a pick-me-up tonic after winter. Cooked in a similar way to spinach, it is only the young leaves that are eaten, as older leaves are bitter and contain calcium oxalate crystals that make them very gritty. The young leaves should not be eaten raw as they will sting.

Other uses
Nettle is an effective natural spray against aphids and other sap suckers. Soak chopped leaves and stems in water for 24 hours, strain and spray. This is also a good tonic spray for plants.

P

PARSLEY
Petroselinum crispum

Growing
- Biennial, hardy
- Fertile soil
- Full sun
- 60 cm high by 80 cm wide

Harvesting
To harvest, snip leaves off several plants rather than denuding one. New growth comes from the middle of the plant, so harvest the outside leaves. Parsley does not dry well, so freezing is a better option. Harvest seeds when dry, and harvest roots at the end of its second year.

Parts used
Leaves, seeds and roots

Uses
- Antiseptic
- Relieves spasms
- Relieves flatulence
- Diuretic
- Nourishing tonic

Cautions
Do not use in more than culinary quantities while pregnant.

I love parsley's Latin name as it sounds exactly like the way parsley tastes! This is one of the most widespread and popular herbs. Even Peter Rabbit knew about eating parsley after overindulging: 'First he ate some lettuce and some broad beans, then some radishes, and then, feeling rather sick, he went to look for some parsley.'

In the garden
Parsley comes in two main varieties: flat leaf (or Italian) and curly leaf. Parsley takes ages to germinate and is easier to grow from seedlings. If you want to sow seeds, soak them in warm water overnight before planting. Although it is a biennial and hardy enough to survive mild frosts, rather plant new seedlings each year as the leaves from the second-year plants tend to become tough and bitter. Leave a few of the second-year plants in the ground and they will start flowering when the heat builds up. These attract many beneficial insects. Parsley likes fertile soil, full sun and plenty of regular moisture. It is a low-maintenance plant once established.

Growing in containers
Parsley is easily grown in containers and is a popular herb to grow on a windowsill. Make sure it is kept evenly watered.

Healing properties
Parsley has a very high vitamin and mineral content, and was used as a medicine long before it became a culinary herb. Internally parsley is used to treat indigestion, rheumatism, colic, water retention, kidney stones and gall bladder inflammation, and to strengthen hair and nails. It is a good general tonic if you are run down or depressed. It will help maintain emotional balance through stress or times of change such as menopause. Externally its antiseptic qualities are used to treat sprains, insect bites and wounds, and as a wash for sore eyes and itchy skin conditions. The root contains high levels of boron and is good to treat bone problems.

How to use
Infusion: (leaves) Internally to treat emotional imbalances (especially during menopause); strengthen hair and nails; help replace iron after breast-feeding or menstruation; reduce inflammation and water retention; treat gall bladder inflammation and kidney stones; to increase breast milk and as a general tonic. Externally as a wash to treat skin conditions such as eczema, psoriasis, pimples and acne.
Decoction: (root) To treat indigestion and colic, and to strengthen bones.
Tincture: (leaves) As for infusion, but stronger.
Compress: (leaves) For sprains, swollen joints, wounds, insect bites and sore eyes.

Culinary uses
Italian parsley has a stronger flavour than curly leaf parsley, which is more often used as a garnish. This is a pity, as the curly leaf variety adds subtle flavour to many dishes. With both varieties, add it towards the end of cooking so its flavour isn't lost. If you want to cook it for longer, use the stems. Parsley is a good addition chopped into egg dishes, salads and sauces or puréed to add to a spice rub or mix.

Other uses
An infusion of crushed parsley seeds will get rid of head lice. Wash the hair first, then dunk the scalp into the infusion. Wrap the head in a hot towel for 30 minutes and then leave to dry.

An old English superstition claims that parsley will grow only if the woman is the head of the house

PELARGONIUM
Pelargonium species

From nutmeg to rose, peppermint to citronella, scented pelargoniums (often incorrectly called geraniums) are indigenous to South Africa.

Growing
- Semi-hardy perennial
- Well-drained soil
- Full sun to semi shade
- 15 cm – 1.3 m high by up to 1.3 m wide

Harvesting
Pick leaves and flowers when needed. The leaves dry well.

Parts used
Leaves and flowers

Uses
- Antidepressant
- Tonic
- Antiseptic
- Pain reliever
- Diuretic
- Sedative

Cautions
As it affects the hormonal system, do not over-use during pregnancy.

In the garden
Pelargoniums like a sunny position with well-drained soil. They do not like having very wet feet or being over-watered. They are quite drought tolerant and half hardy but won't survive severe frosts. I take cuttings of my rarer pelargoniums towards the end of summer and pot them up (see page 23). This ensures I still have the plants in spring if I lose them to a winter frost. They are easily propagated from cuttings. In spring, trim plants back to encourage new leafy growth.

Growing in containers
Pelargoniums are one of the easiest herbs to grow in a container. They don't need huge amounts of water and they look and smell great. Repot every couple of years to prevent the plants becoming pot bound.

Healing properties
There are many varieties with different healing properties. Rose-scented pelargonium (*P. graveolens*) is an excellent herb for calming and de-stressing. It helps with insomnia, eases tension (including premenstrual tension) and anxiety and treats gastroenteritis. It has a balancing effect on the hormonal system and also increases circulation and soothes sore throats. Added to face creams, it soothes inflamed skin. It is also a good face cleanser – especially for problem skins. Camphor-scented pelargonium (*P. betulinum*) helps ease sinusitis and bronchial congestion.

How to use
Infusion: (*P. graveolens* leaves) Internally to treat gastroenteritis and to ease tension, anxiety and insomnia. Externally as a gargle for sore throats, as a face wash for skin conditions and a wash for minor wounds.
Infusion: (*P. betulinum* leaves) Inhale the steam to ease sinusitis and bronchial congestion.
Essential oil: (*P. graveolens*) Added to massage oil to treat depression, insomnia, anxiety and premenstrual tension. Added to face creams to treat skin conditions.
Infused oil: (*P. graveolens* leaves) As for essential oil.

Culinary uses
Similar to bay leaves, pelargonium leaves are better suited to infusing dishes with their flavour, rather than being eaten themselves. The sweetly scented pelargoniums, such as rose and apple, are particularly delicious infused in teas, jellies and cordials. Or use the leaves to infuse warm cream or milk and make custards, ice cream and sauces. To add flavour to baked cakes and muffins, line the baking tin with fresh leaves. Chop leaves and blend with sugar to use in tea or to sprinkle on fruit. The edible flowers can be sugared for cake decorations, frozen in ice cubes or added to salads and desserts.

Other uses
The more strongly scented pelargoniums, particularly citronella, keep harmful insects out of the garden while their pretty pink and white flowers attract beneficial insects. Tear up leaves of the citronella pelargonium and sprinkle over newly seeded beds to repel insects. The leaves repel cabbage moths, making scented pelargoniums beneficial near any of the brassica family. Dried leaves of all scented pelargoniums repel moths from clothes and keep them smelling fresh.

R

ROCKET
Eruca vesicaria

Rocket, also known as arugula, is native to the Mediterranean and is a fast-growing annual with peppery, edible leaves and flowers. Wild rocket has smaller, spikier leaves and is hotter than the domesticated variety. They are both delicious.

Growing
- Hardy annual
- Most soils
- Full sun to semi shade in hot weather
- 60 cm high by 80 cm wide

Harvesting
Start picking outside leaves off young plants about four weeks after sowing. When it goes to seed the leaves are still edible, although they become hotter and the stems thicker. It does not dry well and is better preserved by freezing.

Parts used
Leaves, flowers and seeds

Uses
- Warming
- Drying
- Mild stimulant
- Tonic
- High in antioxidants and minerals

In the garden
Rocket is ridiculously expensive to buy yet extremely easy to grow. If you leave it to self-sow, you will always have rocket in your garden. It likes full sun, except during the hottest summer months, when it prefers semi shade. It likes moist rich soil and will quickly go to seed on hot, dry soil. Direct seed and keep moist until it germinates. You don't need to thin seedlings out, as the stronger plants will crowd out the weaker ones. In hot weather, when it goes to seed quite quickly, do successive sowings every few weeks to ensure a continual harvest.

It is hardy and does not seem to be bothered by pests or disease. Rocket grows well with lettuces, as they both have the same growing requirements. Pansies seem to grow bigger with larger flowers when grown in amongst rocket. It is a happy companion to spring onions and radishes.

Growing in containers
Rocket will grow in a container as long as it is watered regularly and doesn't become too hot or dry. When it goes to seed, make sure the seeds are sprinkled onto the soil and don't fall outside the pot.

Healing properties
Rocket was supposedly used by the Romans as an aphrodisiac. It is rich in antioxidants and vitamins and is a tonic; and acts as a mild stimulant, and treats coughs. The seeds are used to treat bruises.

How to use
Infusion: (leaves and seeds) As a pick-me-up tonic, to stimulate the digestive system, cleanse the body and to strengthen the immune system.
Syrup: (leaves and seeds) To treat coughs.

Culinary uses
Rocket adds a mustardy bite to salads and is a tasty bed for grilled fish or chicken breasts. It can also be added to stir-fries and Asian soup dishes, particularly if you use older rocket with tougher leaves. Rocket pesto is a delicious way to use the harvest bounty when the rocket really gets going in the garden. The peppery seeds can be crushed and made into mustard or infused in hot oil.

R

ROSEMARY
Rosemarinus officinalis

Growing
- Hardy perennial
- Well-drained soil
- Full sun
- Different varieties grow to different heights

Harvesting
Rosemary benefits from being cut, so harvest it regularly.

Parts used
Leaves and flowers

Uses
- Antibacterial
- Antifungal
- Stimulant
- Antiseptic

Cautions
Never take the oil internally. Large doses of the leaf can be toxic.

When I grew my first herbs and vegetables, I planted them around an existing large rosemary bush. Years later it had became so old and straggly I had to remove it. The following spring, a rosemary branch appeared – growing out of the middle of my vegetable garden wall! I have left it to grow and it is now a quirky feature in my garden.

In the garden
Native to the Mediterranean, rosemary now grows throughout the world. Although its preferred climate is hot and dry, it does well in more temperate zones. There are many varieties, some trailing and unruly, others compact and bushy, with flowers ranging from pale mauve to dark purple. It is difficult to grow from seed, so rather purchase seedlings or propagate from an existing plant, using layering or cuttings (see page 23). Choose the spot to plant it carefully; it is a large perennial and can live up to 20 years. Cut it back in late spring. In frost areas, avoid cutting it back in autumn.

Growing in containers
Rosemary does well in large pots, as long as they have good drainage and are not over-watered.

Healing properties
Rosemary has a multitude of medicinal uses, ranging from treating depression and baldness to improving memory and circulation. It is a good hair tonic in herbal shampoos and conditioners. Externally it is used to treat headaches, as a wash for skin conditions and as an insect repellent. It is an effective gargle and mouthwash. Internally it treats flatulence, stimulates digestion and improves memory and circulation.

How to use
Infusion: (leaves) Internally to ease flatulence, to stimulate digestion, to ease depression and to improve memory and circulation. As a mouthwash and gargle to clear infections and freshen breath. As a wash or added to a bath for skin conditions. Use as a hair rinse for shiny hair.
Essential oil: Mix with a carrier oil and apply to temples to ease a headache. Add to a cream to repel insects, increase circulation and treat eczema, psoriasis and dry skin.
Infused oil: As for essential oil. Add to hair conditioner to encourage hair growth.

Culinary uses
Fresh rosemary tastes much better than dried and adds robust flavour to roasts and sauces. The stems can be used as delicious skewers for meat or vegetables. Try adding stems to a braai for a smoked rosemary flavour.

Other uses
Bees and other beneficial insects love rosemary flowers and harmful bugs don't like its strong-smelling leaves. I find rosemary sprigs very useful for protecting young seedlings. See page 29 for details. Rosemary also makes very good hedging, particularly in a vegetable and herb garden as it helps repel harmful insects.

R

RUE
Ruta graveolens

Growing
- Hardy perennial
- Well-drained soil
- Full sun
- 1.5 m high by 1.5 m wide

Harvesting
Pick fresh leaves year round.

Parts used
Leaves and stems

Uses
- Relieves spasms
- Promotes menstrual flow
- Lowers blood pressure
- Stimulates circulation

Cautions
Preferably use under guidance of a trained herbalist. Do not use if pregnant. Rue is toxic in large doses. Do not take the oil internally. Many people are sensitive to the oil in rue's leaves and develop small blisters or a painful rash if they touch them. This is more prevalent during hot weather. To be on the safe side wear gloves and long sleeves when handling rue.

Rue is a strong-smelling herb – as signified by its Latin name *graveolens*. Sometimes I love its smell and at other times I can't bear it. It has pretty blue-green leaves and yellow flowers, making it an attractive addition to the garden.

In the garden
Rue is another native of the Mediterranean. It is a hardy perennial and it doesn't mind poor soil, as long as it is well drained. It grows best in full sun and is very drought resistant. Because the leaves can cause a rash on some people, it is a good idea to position plants away from a path. Transplant seedlings or take cuttings from existing plants in spring (see page 23). It has a reputation of deterring the growth of many vegetables, so it is a good idea to plant it away from your vegetable garden. Cut plants back in spring and again after flowering ends in summer to prevent them becoming straggly.

Growing in containers
Rue should be grown in a large container as it has quite a deep root. Make sure it is positioned where people can't brush against the leaves.

Healing properties
Rue has been used for hundreds of years as a protective plant, to ward off evil spirits and as an antidote against poison. It is said to improve not only eyesight but also second sight and creativity. Great artists such as Michelangelo and Leonardo da Vinci supposedly claimed that rue's mystical powers improved both their sight and artistic vision.

Rue contains high levels of rutin. When taken internally, rue strengthens blood vessels and capillaries, lowering blood pressure and increasing circulation. As with many bitter herbs, it stimulates bile and aids indigestion. It also stimulates the uterus, easing cramps and inducing suppressed menstrual flow. Externally it helps ease painful nervous conditions such as shingles and neuralgia.

How to use
Infusion: Externally as an eye wash to soothe tired eyes and clear bloodshot eyes. Internally to calm nervous tension, reduce blood pressure, prevent and ease varicose veins, increase circulation and blood flow to the heart, improve eyesight, improve digestion, reduce menstrual cramps, induce menstrual flow and rid the body of intestinal worms.
Ointment: To treat insect bites, fungal skin conditions, varicose veins, gout and rheumatic pains, backache, joint pain, arthritis, shingles, neuralgia and sciatica.
Poultice: To relieve sciatica, shingles, neuralgia, backache and rheumatic pains.
Infused oil: As for ointment and poultice.

Culinary uses
Rue is a bitter-tasting herb and is one of those you either really like or can't stand. Use leaves sparingly in fish or egg dishes and add to cream cheese or vinegar.

Other uses
Rue's strong smell makes it a good insect repellent. Use clippings as a mulch to repel harmful insects or hang them in windows to keep flies and mosquitoes away. Dried leaves can be sprinkled around ant entry points to stop them coming into the house.

S

SAGE
Salvia officinalis

Despite its popularity in the kitchen, sage has been used for far longer as a medicinal herb than as a culinary one. In ancient times, sage symbolised wisdom, good health and long life.

Growing
- Hardy (except for pineapple sage which will not survive severe frost), perennial
- Well-drained soil
- Full sun
- Up to 60 cm high by 60–100 cm wide

Harvesting
Pick leaves throughout the year.

Parts used
Leaves

Uses
- Astringent
- Antiseptic
- Stimulant
- Relieves spasms
- Generally strengthening

Cautions
Do not take if breast-feeding. Do not drink tea medicinally for longer than two weeks, as extended use can be toxic.

In the garden

Sage comes in a variety of colours, sizes and scents, from the deliciously sweet pineapple sage, with its bright red spiky flowers, to purple, tricolour and pale green varieties. Native to the Mediterranean, it needs well-drained soil and plenty of sun. It likes being pruned in early spring, which not only keeps it healthy but also maintains its shape. After five years or so, sage becomes woody and straggly and needs to be replaced. Grow new plants from the parent plant by layering (see page 23).

Growing in containers

Sage grows well in a well-drained container. Do not over-water it.

Healing properties

The Latin name of this herb means 'to heal' or 'to save'. Egyptian women drank sage to increase fertility and the Chinese revered it so much they were willing to trade four times the weight of fine tea for sage. It is an astringent and antiseptic, making it very useful for mouthwashes and healing sore throats.

How to use

Infusion: To treat coughs and colds, weak lungs, allergies and persistent coughs. To lift depression and ease nervous exhaustion (especially in the elderly). To ease indigestion, particularly after eating fatty food. As a gargle for sore throats and laryngitis and as a mouthwash for ulcers, mouth infections and inflamed gums. (Combine with apple cider vinegar for extra effect.) As a wash for slow-healing wounds.

Tincture: To treat night sweats and hot flushes.

Culinary uses

Sage is a strong herb and should be used sparingly. It combines well with fatty foods such as pork. It has traditionally been used in sausages because it helps preserve them. Its leaves are a good addition to flavoured herb vinegars or oils. Try flavouring apple jelly with sage leaves as a delicious condiment for roast pork. Sage butter – a handful of leaves cooked in butter until they are crisp and the butter is nutty – is delicious with gnocchi or ravioli.

Other uses

Sage is a valuable companion plant with its spiky flowers attracting hordes of beneficial insects and its strongly scented leaves repelling many of the bad guys. It is particularly beneficial planted near any members of the cabbage family. Scatter trimmings as mulch to confuse birds and repel insects.

S

SALAD BURNET
Sanguisorba minor

Growing
- Hardy perennial
- Well-drained soil
- Full sun to semi shade
- 40–50 cm high by 60 cm wide

Harvesting
It does not dry well so pick fresh leaves when needed.

Parts used
Leaves and roots

Uses
- Cooling
- Astringent
- Stops bleeding
- Antibacterial
- General healing

Francis Bacon once said of salad burnet that it should be grown in pathways along with thyme and water mint 'to perfume the air most delightfully, being trodden on and crushed'.

In the garden
This hardy perennial bears little greenish-red pompom flowers from late spring to midsummer. It likes well-drained soil and is drought tolerant. It grows in full sun to semi shade. If you want to encourage bushier growth, give it a trim in early spring to prevent it flowering. However, if you leave it to flower it attracts beneficial insects such as hoverflies and bees. Propagate by dividing the plants in spring or autumn (see page 25).

Growing in containers
Salad burnet grows well in containers, providing green all year round. Do not over feed otherwise it loses its flavour.

Healing properties
Salad burnet's Latin name means 'blood absorb' and its traditional use was to staunch bleeding. During the American Revolution, soldiers drank salad burnet tea in the belief that any wounds received would be less severe. The roots contain high levels of tannin, which helps to stop bleeding. A poultice made from the roots was used to staunch bleeding wounds, and reduce haemorrhaging and internal bleeding. Salad burnet soothes sunburn, insect bites and inflamed skin conditions. It is a calming and uplifting herb.

How to use
Infusion: (leaves) Externally as a cooling wash for sunburn. Internally to lift depression.
Decoction: (roots) Internally to reduce bleeding and nosebleeds.
Ointment: (leaves) To treat inflamed skin conditions such as eczema and to soothe burns, wounds, insect bites and stings.
Poultice: (roots) To help stop wounds from bleeding.

Culinary uses
Salad burnet has a nutty taste and, as its name implies, it can be added to salads and chopped up into salad dressings, especially in winter. Use younger leaves, as older ones can become bitter. Leaves are a refreshing addition to fruity drinks and can be added to cream cheese, herbal vinegar and herb butters.

SAVORY – WINTER AND SUMMER

Satureja montana and *Satureja hortensis*

Growing
- Winter savory: hardy perennial
- 30 cm high by 30 cm wide
- Summer savory: tender annual
- 45 cm high by 20 cm wide
- Well-drained soil
- Full sun

Harvesting
Pick winter savory when needed. Pick summer savory before it flowers to dry for winter use.

Parts used
Leaves

Uses
- Relieves flatulence
- Antiseptic
- Astringent

Cautions
Do not take medicinally if pregnant.

Savory's Latin name is derived from 'satyr', the half-goat, half-man creatures with insatiable sexual appetites. According to legend, satyrs lived in meadows of savory; so no wonder this herb acquired the reputation of being an aphrodisiac.

In the garden
The hardy perennial winter savory is a low-growing, sprawling plant. It produces fragrant white to lilac flowers that attract bees. Trim winter savory in spring to encourage new growth.

Summer savory is a more delicate tender annual that will reach a height of about half a metre. It has tiny white to pink flowers in late summer. The pale green, slender leaves grow sparsely along delicate, rust-coloured stems. It can become leggy if the leaves aren't picked regularly. Both savories like well-drained soil in full sun.

Growing in containers
Both varieties of savory grow in well-drained containers. Avoid over-watering winter savory.

Healing properties
Both varieties stimulate appetite and are a remedy for flatulence, indigestion and digestive upsets. Their fragrant oils have antiseptic and astringent properties, effective for treating sore throats and colds.

How to use
Infusion: To stimulate appetite; to treat colic, flatulence, cystitis, bronchial congestion, sore throats and colds.
Tincture: As for infusion, but stronger.

Culinary uses
The leaves of winter savory are quite tough and are best added to dishes requiring long cooking. The heat and moisture break them down, releasing flavourful oils and making them palatable. Cooked with beans and lentils, winter savory aids digestion and reduces flatulence. It is a particularly good herb to use with Jerusalem artichokes.

Summer savory adds a flavourful kick to fresh green beans. The leaves are very tender and can be eaten in salads or added to vinegar, giving it a fresh summer taste.

Other uses
In the garden, both varieties of savory repel leaf-eating beetles.

SELFHEAL
Prunella vulgaris

Growing
- Hardy perennial
- Well-drained, fertile soil
- Full sun to semi shade
- 20–30 cm high and spreading

Harvesting
Harvest the leaves and flowers throughout the growing season. They dry well for later use.

Parts used
Leaves and flowers

Uses
- Bitter
- Cooling
- Antibacterial
- Astringent
- Reduces blood pressure
- Stops bleeding
- Liver stimulant

There is an old German proverb that goes: 'He needeth neither physician nor surgeon that hath selfheal to help himself.' Selfheal – as its name suggests – has been used for centuries to treat a wide range of ailments. This spreading little herb is also known as the carpenters' herb because of its ability to stop wounds from bleeding.

In the garden
Selfheal is an easy-to-grow, creeping perennial and will spread to form quite a dense ground cover. From summer to autumn it produces spikes of purple flowers loved by bees and butterflies. It is easily propagated by division off an existing plant (see page 25). It likes full sun to semi shade and prefers fertile, well-drained soil.

Growing in containers
Selfheal grows well in containers combined with taller plants.

Healing properties
Selfheal stops bleeding very effectively. It also treats mouth ulcers, gum infections, cuts and wounds, sore throats and tired eyes. In Chinese medicine the flower spires are used to treat what is called 'liver-fire', the result of a sluggish liver leading to agitation, anger, dizziness, insomnia and headaches. The English phrase 'gung ho' is derived from the Chinese term for this condition – *gan hao*. And selfheal cools down a person who has excessive 'gung ho'.

How to use
Infusion: (leaves and flowers) Internally and externally, as for tincture. A weak infusion as an eye wash for hot, tired or infected eyes.
Decoction: (flowers) To treat high blood pressure, anger, hyperactivity and irritability associated with a sluggish liver.
Tincture: (leaves and flowers) Internally for all bleeding, including nosebleeds or heavy menstruation. As a gargle for sore throats and mouthwash for mouth infections.
Poultice: (fresh leaves) Applied to fresh wounds to stop bleeding and to encourage healing. Place in nostrils to stop a nosebleed.

SOAPWORT
Saponaria officinalis

Growing
- Hardy perennial
- Well-drained, poor soil
- Full sun to semi shade
- 30–90 cm high and spreading

Harvesting
Pick leaves when needed. Harvest roots in autumn and dry for later use.

Parts used
Leaves and roots

Uses
- Relieves inflammation
- Soothing

Cautions
This herb should be taken internally only under guidance of a trained herbalist as its saponin content can be poisonous in high doses.

As its common and Latin names suggest, this plant makes an effective soap. It contains high levels of saponins, which foam when mixed with water. It is also a pretty addition to the garden with its masses of pink flowers.

In the garden
Soapwort is a perennial and quite a vigorous spreader. It grows easily from cuttings (see page 23) or seedlings. It prefers full sun but will grow in semi shade. Plant it in well-drained, poor soil. It can become invasive if it is in rich soil. Do not grow near a fishpond as the saponins can harm fish. It will die back in winter in cooler areas but pop up again in spring.

Growing in containers
Soapwort grows well in containers and this will prevent it from spreading.

Healing properties
Soapwort is a soothing and anti-inflammatory plant, effective on itchy and dry skin conditions. It is also good for reducing swelling related to rheumatism and gout.

How to use
Infusion: (leaves and roots) As the base for conditioning shampoo and body wash; as a wash for rashes, itchy dry skin, psoriasis, eczema, boils and acne.
Ointment: (leaves and roots) To treat rashes, itchy dry skin, psoriasis, eczema, boils and acne. Rub on inflamed joints to reduce swelling.

Other uses
Use an infusion of soapwort in the garden as a spray against aphids and other sap-sucking insects. It makes a good wash for delicate garments.

S

SORREL
Rumex acetosa

Sorrel's name comes from the Teutonic word for sour, giving an idea of this herb's essential flavour. It was used by the Romans to aid digestion, particularly after a fatty feast.

Growing
- Hardy perennial
- Well-drained soil, will handle poor soil
- Full sun to semi shade
- 60–100 cm high by 60–80 cm wide

Harvesting
Regular harvesting will keep the leaves tender. Pick leaves throughout the year. Sorrel leaves do not dry well.

Parts used
Leaves

Uses
- Tonic
- Digestive stimulant
- Diuretic
- Blood cleanser

Cautions
Do not take in excess as its high levels of oxalic acid can cause kidney damage. It should not be taken medicinally if you suffer from gout, rheumatism, arthritis, kidney stones or gastric hyperacidity.

In the garden
There are two varieties: French sorrel, with green leaves, and the milder bloody sorrel, with striking red-veined leaves. It is a hardy perennial and likes fertile soil in full sun to semi shade. Once plants are established, divide them in autumn to keep the leaves tender (see page 25). In late summer it forms tall spikes of flowers. If you let it go to seed you will be pulling sorrel out from everywhere in the months to come, so keep cutting flowers off. This encourages leafy growth to continue. In winter it will die back in colder areas.

Growing in containers
Sorrel does well in a wide container that allows it room to spread.

Healing properties
Sorrel is a blood cleanser and tonic. Its bitter flavour encourages bile formation, making it a good digestive herb, stimulating appetite and aiding digestion. It has high levels of vitamin C.

How to use
Infusion: Internally as a tonic, digestive stimulant, blood cleanser and diuretic. Externally as a wash for acne, pimples and as a skin cleanser.

Culinary uses
Sorrel has a tangy, lemony flavour, adding a delicious bite to salads. It can be used as a flavouring herb or as a leafy green. Its flavour is stronger in the middle of summer when larger, older leaves can be cooked.

Other uses
Juice from the sorrel leaf will remove mould and rust stains from fabric. It will also clean stained silver.

Blend sorrel and parsley together with olive oil, pine nuts and Parmesan cheese to make a delicious lemony-tasting pesto

S

ST JOHN'S WORT
Hypericum perforatum

Growing
- Hardy perennial
- Well-drained, poor soil
- Full sun to semi shade
- 30–90 cm high by 30–40 cm wide

Harvesting
Pick leaves just before flowering. Pick flowers in early summer, just as the buds have opened. The herb dries well.

Parts used
Leaves and flowers

Uses
- Bitter
- Cooling
- Drying
- Antidepressant
- Antibiotic
- Relieves inflammation
- Sedative
- Astringent
- Restorative to nervous system

This herb has long been used as a protective plant, warding off everything from evil spirits and sickness to lightning. It was once carried to increase courage and endurance. So the next time you run a marathon – take some sprigs of St John's wort with you.

In the garden

If you plan to grow St John's wort to use medicinally, make sure you plant *Hypericum perforatum*. There are ornamental varieties that have no healing qualities. The Latin name *perforatum* means perforated and you can tell why by looking at a leaf. Hold one up to the light and you will see what looks like tiny holes in it. These are translucent oil glands, which distinguish the medicinal variety from others.

This pretty herb, with bright yellow flowers, likes full sun or semi shade. It doesn't mind a rich soil but prefers a light soil and can become invasive if the conditions are right. Cut it back after flowering to prevent it self-seeding. In autumn, divide plants (see page 25). In spring, cut back any dried stems to encourage new growth.

Growing in containers

St John's wort grows happily in containers.

Healing properties

St John's wort is used so extensively to treat anxiety and depression that it has earned the nickname 'natural Prozac'. In addition to being an antidepressant it is also an antibiotic and anti-inflammatory. Internally it is an effective treatment for painful joint and nerve conditions. Externally, it is used as a treatment for wounds, burns, painful nerve conditions and swollen joints. Recent studies have shown that it has effective antiviral properties.

How to use

Infusion: (leaves and flowers) Internally to treat anxiety, depression, neuralgia, nervous tension and irritability. Helps treat menopausal mood swings and premenstrual tension. Externally as a wash to treat minor skin wounds and burns.
Tincture: (leaves and flowers) As for infusion, but stronger.
Cream: (flowers) To treat sciatica, shingles, neuralgia, cramps, sprains, cold sores, stings, sores and grazes.
Infused oil: (flowers) Externally, apply to burns or painful sciatica, shingles, neuralgia, cold sores and inflamed joints.
Extract: (flowers) Internally to treat depression, anxiety and nervous tension. Helps treat menopausal mood swings and premenstrual tension.

Cautions

It is recommended that you take the herbal extract internally only under supervision of a qualified herbalist. If you have livestock, keep them away from St John's wort as it can cause poisoning. After taking internally, some people become more sensitive to sunlight, which may cause a skin rash. Be cautious when handling the plant, particularly in hot, wet conditions, as this too can cause a rash. Do not use if pregnant or breast-feeding.

S

STEVIA
Stevia rebaudiana

Growing
- Tender, perennial
- Well-drained, moist soil
- Full sun to semi shade
- 45 cm high by 45 cm wide

Harvesting
Pick leaves fresh during spring and summer. They retain their sweetness after drying.

Parts used
Leaves

Uses
- Sweetening
- Digestive
- Regulates blood sugar
- Reduces high blood pressure

Cautions
The medicinal effects of this herb are still being researched.

A native of South and Central America, stevia contains a substance called stevioside, which is 300 times sweeter than sugar without its fattening or glucose-altering effects.

In the garden
In colder climates this tender herb is grown as an annual. In warmer areas, it can be grown as a perennial. It likes full sun to semi shade, and moist, light soil. It bears white flowers in late summer to early autumn. Cut these to encourage bushier growth. If grown as a perennial, give it a trim in spring to encourage new growth.

Growing in containers
Stevia grows well in containers and can be brought indoors in frost areas to over-winter in a warmer area.

Healing properties
The Guarani Indians of Paraguay used stevia extensively as a sweetener and as a remedy for indigestion. However, it has only recently gained prominence as a modern herbal remedy. Scientific research is showing it is beneficial in regulating blood sugar levels, reducing obesity and as a treatment for high blood pressure. It is a healthy alternative to chemical sweeteners. It is also used to aid digestion and suppress bacterial decay around teeth.

How to use
Infusion: To ease indigestion, regulate blood sugar and reduce high blood pressure. As a mouthwash to suppress tooth decay.

Culinary uses
Stevia leaves are intensely sweet with an almost liquorice taste. After they are dried, they can be crushed to a fine powder and used as a sweetener. The sweetness can also be extracted by adding fresh leaves to just boiled water and leaving them to steep for 10 minutes. This water can be used to sweeten drinks or as a base for a dessert. I often add a stevia leaf to a bitter herbal tea to make it more palatable.

The leaves do have a slightly bitter aftertaste, which comes from the veins of the leaf. Most commercial stevia powder, available from health shops, is made using extraction methods that reduce this bitterness. Stevia powder can be used as a sugar substitute but don't make the mistake of using the same amount of stevia as you would sugar – one teaspoon of stevia powder is equal to about one cup of sugar!

T

TANSY
Tanacetum vulgare

Growing
- Hardy perennial
- Well-drained, fertile to poor soil
- Full sun
- 90 cm high by 60 cm wide and spreading

Harvesting
Pick leaves as and when needed.

Parts used
Leaves

Uses
- Expels worms
- Stimulant

Cautions
Do not take internally if pregnant. Use tansy internally only under the supervision of a trained herbalist. Large doses can lead to seizures and death.

Tansy is an excellent companion plant in a vegetable and herb garden. Despite it repelling many harmful insects I recently discovered that it is a Mecca plant for ladybirds. Judging by their behaviour in my garden they find tansy an aphrodisiac!

In the garden
Tansy, with its bright yellow button flowers, grows in most soils and likes full sun. It doesn't like having wet feet, so make sure the soil is well drained and you don't over-water it. It grows into quite a bushy, straggly plant by sending out vigorous rhizomes, and can be invasive. Keep it in check by continually cutting it back and pulling out the new growth around the edges of the main plant.

It has such good insect-repelling properties it was once used to embalm bodies and slow decomposition. Its pungent leaves deter everything from beetles and bugs to flies, ants and mice. Use the clippings as a pest deterrent by scattering them on newly seeded ground and sprinkling them amongst young seedlings. Tansy will self-seed in the right conditions. Control this by cutting off the flowers before they set seed. In cold areas it will die back during winter, but in spring it will shoot up again.

Growing in containers
Because it can become invasive it is a good idea to keep tansy contained in a pot. Make sure it is well drained and not over-watered.

Healing properties
Tansy is an effective treatment for roundworm and tapeworm – however any internal treatment using this herb should be done under guidance of a trained herbalist. Externally it can be used to treat rheumatism, migraine and neuralgia, and skin conditions such as scabies.

How to use
Infusion: Externally to treat scabies.
Compress: To treat painful rheumatism, migraine and neuralgia.

Other uses
Tansy has a high potassium content and is a good addition to the compost heap. Hang sprays of tansy in windows to keep flies away. Sprinkle dried leaves down ant holes or at their entrance points to keep ants out of the house.

T

TARRAGON
Artemisia dracunculus

Tarragon's Latin name means 'little dragon', which dates back to the belief that this herb was an antidote for the bites of poisonous animals.

Growing
- Semi-hardy perennial
- Well-drained soil
- Full sun to semi shade
- 90 cm high by 45 cm wide

Harvesting
Pick young sprigs in spring and early summer to preserve in vinegar. For fresh use, pick throughout the growing season. For drying or freezing, pick in midsummer. Leaves retain flavour when dried or frozen.

Parts used
Leaves

Uses
- Digestive
- Calming
- Pain reliever

In the garden
There are two closely related types of tarragon: French and Russian. French tarragon, with its delicate aniseed flavour, is the preferred herb. It is harder to find seedlings of this variety, as they do not produce viable seed. Tarragon is a semi-hardy perennial, which dies down during winter. It likes full sun to semi shade and soil that is not too rich, but it must be well drained as tarragon does not like wet feet. In summer, remove any flowers to encourage leaves to grow. In winter, mulch well with a thick layer of compost to protect the plant while dormant. Older plants tend to lose their flavour. To keep a constant supply of young plants, dig up some runners every spring and pull them apart. Pot up sections that have growth nodules and plant them out once they are established.

Growing in containers
French tarragon grows well in containers. It produces runners, so choose a large pot with room for the plant to spread.

Healing properties
Although it is almost solely used for culinary purposes today, tarragon does have medicinal properties. It does more than just flavour a rich sauce – it also contains enzymes that aid digestion, especially after eating protein. It is a calming herb, easing insomnia. Fresh leaves help anaesthetise painful toothache or cuts and sores. It is full of vitamins and minerals and is a good all-round tonic.

How to use
Infusion: (leaves) To treat insomnia; stimulate appetite and ease indigestion; as a general tonic.
Whole leaves: Chew to freshen breath or to ease painful toothache.

Culinary uses
Although tarragon is subtle, it infuses dishes very quickly with its flavour and should not be used excessively. Fresh tarragon leaves are ideal for flavouring egg dishes, chicken, sauces, salad dressings and soups. Tarragon vinegar, made using good-quality white wine vinegar, is a tasty way to preserve leaves and produce flavoured vinegar.

Other uses
Because it is high in minerals tarragon is a good addition to the compost pile, helping break it down and adding valuable nutrients.

T

TEA TREE
Melaleuca alternifolia

Aboriginal Australians have been using the sap from the tea tree for centuries as a potent all-purpose healer.

Growing
- Hardy perennial
- Well-drained, moist soil
- Full sun
- 7 m high by 3 m wide

Harvesting
Cut leaves from spring to midsummer.

Parts used
Leaves and branches

Uses
- Antimicrobial
- Antifungal
- Antibacterial
- Antiviral

Cautions
Do not take internally. Do not use if pregnant or breast-feeding.

In the garden
Tea tree grows up to 7 m tall but it can be kept trimmed smaller. It is a graceful tree with peeling white paper bark, light spiky foliage and creamy, feathery flowers in summer. Although it prefers moist soil, it will grow in most soils, as long as it is well drained. It is hardy and likes a sunny spot. If you are keeping it trimmed, cut it back after it has finished flowering. Keep it well watered throughout summer. It is hardy but when young will need protection in areas with severe frosts.

Growing in containers
Tea tree will grow in a large container but needs to be kept trimmed. Make sure it is kept well watered.

Healing properties
When Captain Cook and his crew landed on the shores of Australia, the Aborigines taught them how to treat their cuts using the crushed leaves from the tea tree. The essential oil from the tea tree has powerful antimicrobial, antifungal, antibacterial and antiviral properties. It is used to treat fungal infections, acne, mouth and gum infections and to disinfect cuts, sores and rashes. It is also an effective inhalant remedy for coughs and colds.

Aboriginal Australians used tea tree bark as burial shrouds

How to use
Infusion: Apply to fungal infections. As a mouthwash and gargle for mouth infections, sore gums, ulcers and throat infections. As a douche for thrush. Use to clean and disinfect sores, cuts and rashes. As a face wash for pimples and acne.
Steam inhalation: Either use a hot infusion or add infused or essential oil to boiling water and inhale the steam to treat coughs and congested colds.
Ointment: To treat acne, cuts and sores.
Essential oil: Apply to acne, leave overnight (see also page 218). Add a few drops to apple cider vinegar or water and use as a mouthwash and gargle for mouth infections, sore gums, ulcers and throat infections.
Infused oil: As for essential oil. Rub onto fungal skin conditions and sores and cuts.

Other uses
Use tea tree oil mixed with vinegar or water to disinfect kitchen and bathroom surfaces. (See page 212 for a recipe.) Add a few drops of oil to body cream and use as a mosquito repellent.

THYME
Thymus species

Growing
- Hardy perennial
- Well-drained, light soil
- Full sun
- Varies from upright and bushy to low-growing and spreading.

Harvesting
Cut stems rather than tugging, as this can pull the roots up. Leaves can be harvested year round, but are at their most flavoursome just before flowering. Ordinary thyme dries very well, however the lemon varieties lose their flavour and are best used fresh. Thyme leaves keep well in vinegar or oil.

Parts used
Leaves and flowers

Uses
- Warming
- Disinfectant
- Antibacterial
- Antifungal
- Antiseptic
- Expels phlegm
- Astringent
- Antimicrobial
- Soothes coughs
- Antibiotic
- Heals wounds
- Stimulates circulation
- Immune-boosting

'I know a bank where the wild thyme blows, where oxlips and the nodding violet grows,' says Oberon, King of the Fairies in Shakespeare's *A Midsummer Night's Dream*. He is describing where the Fairy Queen Titania sleeps because, as all savvy gardeners in Shakespeare's time knew, if you want to invite fairies into your garden – plant a bed of thyme.

In the garden
Thyme is a valuable plant in the garden. Hardy and unfussy, it is easy to grow, is beneficial to other plants and tastes good. For a small plant, it packs a big punch. It is a perennial, bearing pretty pink and purple flowers in summer. There are numerous varieties, including variegated and lemon-flavoured ones. Originating in the Mediterranean, it does well in hot, dry climates. However, it grows happily through rainy summers, as long as it has sufficient sunlight and the soil is well drained.

It can be grown from seed, but is much easier from seedlings or root divisions off an existing plant. It doesn't like rich soil, preferring light earth. Every year after flowering, give the plant a trim to encourage new growth and prevent it from becoming too woody and sprawling. After about three years, the plant will become 'leggy' and should be divided. Dig it up and clear away some soil. Use a spade to divide it into smaller pieces, each with a section of root and foliage, and replant.

Growing in containers
Thyme loves growing in containers. Creeping thyme covers the surface of a pot and spills attractively over the sides.

Healing properties
For centuries thyme has been used for a wide range of purposes. Ancient Egyptians used the herb to mummify bodies. The Greeks burnt thyme as sacred incense, and marauding Scottish highlanders, brave knights of the Middle Ages and Roman soldiers all believed that eating thyme would increase their strength and courage. The essential oil of thyme – thymol – was isolated by the German apothecary Neuiuiann in 1725. Thymol is a strong antiseptic and effective cough remedy. It is good for loosening stubborn phlegm, strengthening lungs and relieving hay fever. Thyme is a digestive aid and helps break down fatty food. In addition, thyme has a multitude of other uses: it is astringent, antibacterial, antifungal, antimicrobial, antibiotic, and stimulates circulation. It contains a wide variety of antioxidants, vitamins and minerals, making it a healthy and immune-boosting herb.

How to use
Infusion: (leaves and flowers) To treat asthma, bronchial infections and hay fever. As a gargle for sore throats or a mouthwash for infected gums, mouth ulcers or abscesses. To boost the immune system and to stimulate circulation. Externally as a wash to treat fungal infections such as thrush and athlete's foot.
Tincture: (leaves and flowers) To treat diarrhoea, as an expectorant or diluted as a gargle for sore throats, or mouthwash for infected gums or mouth ulcers.
Syrup: (leaves and flowers) As an expectorant and to treat bronchial conditions.
Poultice: (leaves) To disinfect minor wounds and soothe insect stings.
Essential oil: Added to a chest rub for bronchial infections. Use diluted on infected sores, insect bites or stings. Add a few drops to your bath or massage oil to ease rheumatism.
Infused oil: (leaves and flowers) As for essential oil, but less diluted.

Culinary uses
The best culinary varieties are common and lemon thyme. It is a robustly flavoured herb and does well in slow-cooked stews and roasts. It is an ideal companion for both pork and lamb. Thyme is a traditional ingredient in *bouquet garni*, along with marjoram, parsley and bay.

Other uses
Thymol is also an insect repellent, making it a handy companion plant. Creeping thyme is a valuable groundcover in a vegetable garden. It spreads quickly, providing a thick, moisture-conserving and insect-repelling layer. Thyme spray can be used to repel insects, in both the garden and the kitchen (see page 259 for a recipe).

Cautions
Thyme oil can irritate mucous membranes, so always dilute before using. Avoid medicinal use of thyme while pregnant, as it is a uterine stimulant. Do not use the essential oil when breast-feeding.

VIETNAMESE CORIANDER
Persicaria odorata

Growing
- Semi-hardy perennial
- Well-drained, moist soil
- Full sun to semi shade
- 45 cm high and spreading

Harvesting
Pick leaves when needed.

Parts used
Leaves

Uses
- Digestive

I first tasted this herb in Malaysia, where it was served in a delicious spicy noodle soup called *laksa*. Also known as Vietnamese mint or cilantro, it is a common ingredient in many Southeast Asian dishes.

In the garden
Vietnamese coriander grows easily from seedlings and cuttings (see page 23). It likes full sun to semi shade and moist, well-drained soil. As it is a tropical plant it does not survive a hard frost. In milder frosts the leaves will die back but it will spring up again in warmer weather. In warm areas it can become invasive, so make sure you keep it trimmed.

Growing in containers
Vietnamese coriander grows happily in containers in a well-drained growing medium. Don't let it get too dry.

Healing properties
It is primarily used as a digestive, easing indigestion, flatulence and colic.

How to use
Infusion: Drink to settle any stomach problems.

Culinary uses
Vietnamese coriander is best added as a garnish at the end of cooking as heat reduces its flavour. It is an essential addition to laksa soup and it adds a zing to many Asian coconut curries or stir-fried vegetable dishes.

W

WATERCRESS
Nasturtium officinale

Growing
- Hardy, short-lived perennial
- In water or very moist soil
- Semi shade
- 50–120 cm high and spreading

Harvesting
Pick young leaves before the plant flowers as it becomes more bitter after flowering. Leaves dry well for medicinal use.

Parts used
Leaves

Uses
- Blood cleanser
- Reduces fever
- Metabolic stimulant
- Detoxifier
- Expels phlegm
- Antioxidant
- Cellular regenerator
- Diuretic
- Expels worms

Cautions
Watercress can cause gastric upsets, especially in people with a history of gastrointestinal ulcers.

A native of central Asia and Europe, watercress is one of the oldest leafy greens eaten by humans, dating back 3 000 years to the ancient Greeks, Romans and Persians. Hippocrates, who understood the connection between what we eat and our health, once said, 'Let food be thy medicine.' In 400 BC he grew wild watercress in the natural springs outside his hospital and used it effectively to treat his patients.

In the garden
Watercress will grow only if its roots are in water or very moist soil. If you have a pond, direct sow the seed along the edges and it will spread happily. Even easier than growing it from seed is buying a bunch of watercress and popping the cut ends into a bottle of water. Put it on a sunny windowsill and change the water every day until roots start growing. Watercress likes semi shade and especially likes being shaded from hot afternoon summer sun.

Growing in containers
Watercress can be grown in clay pots standing in a tray of water. Change the water every day or so to keep it fresh.

Healing properties
Watercress is a member of the cabbage family. Similar to mustard and horseradish, it contains mustard compounds, giving it a distinctive peppery taste and much of its medicinal value. These pungent antibacterial and antifungal compounds help fight colds, mouth infections, flu and bronchitis. Watercress contains many minerals, including iodine, iron and phosphorus, plus high levels of vitamin C and other vitamins (including A, B1, B6, K and E), more iron than spinach and more calcium than milk. In addition, it contains beta-carotene and other antioxidants, making it nature's ultimate multivitamin plant. It is a tonic plant with detoxifying abilities, cleansing the blood, toning the liver and ridding the body of toxins. It also improves sluggish indigestion and clears skin problems. Its high calcium content strengthens bones and teeth.

How to use
Infusion: (fresh or dried leaves) To detoxify and cleanse the body (especially the bloodstream), improve digestion and treat gastrointestinal disorders. (CONTINUED OVERLEAF)

Watercress is a delicious superfood

It tones the liver, clears and improves complexion, strengthens bones and teeth and reduces the risk of lung cancer (particularly in smokers) and prostate cancer. To treat congested chests and bronchial conditions, as an expectorant for phlegmy coughs, to help treat anaemia and as a nutritious drink. Externally for skin problems and to prevent hair loss. As a gargle for mouth and gum infections.

Tincture: (fresh leaves) Diluted internally or added to an ointment to treat previously mentioned skin conditions.

Juice: (fresh leaves) To treat acne, fungal infections of the skin, rashes, and skin infections and irritations.

Poultice: (fresh leaves) To treat mouth ulcers, boils, abscesses, acne and other skin problems.

Culinary uses

In Latin, watercress means 'nose twister', which makes sense when one considers its biting fresh taste. It can be eaten raw in salads, sandwiches and wraps, cooked until just wilted in soups and stir-fries, made into a tasty pesto, added to quiches and muffins, chopped into egg dishes or blended into a healthy juice or cocktail. Once you start growing your own, it will produce seeds. Grind these to make fresh peppery mustard.

Other uses

Watercress flowers are a rich source of pollen and attract bees and other pollinators to your garden.

W

WILD GARLIC
Tulbaghia violacea

Growing
- Hardy perennial
- Well-drained, fertile soil
- Full sun to semi shade
- 30–40 cm high by 40 cm wide

Harvesting
Pick flowers and leaves when needed. Harvest bulbs from mature plants.

Parts used
Leaves, bulb and flowers

Uses
- Antiseptic
- Expels phlegm

This popular member of the onion family is a wonderfully easy plant to grow. Indigenous to southern Africa, it is used to treat a variety of ailments.

In the garden
Wild garlic grows easily from seedlings and from dividing existing plants (see page 25). It is a drought-hardy, clump-forming perennial, which will quickly fill a bed. It likes full sun but can grow in semi shade, however it won't flower as well. It likes well-drained, fertile soil and not too much water. It flowers on and off throughout summer. It is a good companion plant to have in the garden as its strong-smelling leaves are insect-repelling and the flowers attract butterflies and bees.

Growing in containers
Wild garlic grows happily in a container. Don't over-water.

Healing properties
Like many members of the onion family, wild garlic is an antiseptic and expectorant.

How to use
Infusion: (leaves and flowers) Internally to treat coughs, colds, asthma and bronchial infections.
Decoction: (bulbs) To treat coughs and colds.

Culinary uses
The leaves of wild garlic can be used in the same way as chives. The bulbs can be chopped and eaten raw or added to a dish towards the end of cooking, similar to a spring onion. Add the flowers to salads or mix with butter to give it an interesting flavour and colour.

Y

YARROW
Achillea millefolium

Yarrow has long been associated with witchcraft and used as protection against evil. The famous 'stalks of divination' required for consulting the Chinese oracle, the *I Ching*, were made of yarrow. More prosaically, the Druids used yarrow stems to forecast the weather.

Growing
- Hardy perennial
- Most soils
- Full sun
- 30–90 cm high by 60 cm wide and spreading

Harvesting
Harvest leaves throughout the season. Harvest flowers when the plant is in full flower. Cut the leaves from the stems and dry separately. The leaves are very fine and need to be dried quickly. Leave the flowers on the stems and hang until dry, then cut the flowers off and store separately.

Parts used
Leaves and flowers

Uses
- Stops bleeding
- Astringent
- Reduces fever
- Relaxes blood vessels
- Stimulates digestion
- Relieves inflammation
- Anti-allergenic
- Relieves spasms

In the garden

Yarrow is of immense value in the garden because it accumulates nutrients and recycles them back into the soil, improving its quality and benefiting all nearby plants. Yarrow planted in a medicinal herb garden will increase the efficacy of the nearby herbs. Its leaves repel many harmful insects, while its delicate flowers attract many beneficial ones. And, finally, yarrow is of great benefit to compost, speeding up decomposition.

It is a hardy plant, which likes full sun. Other than needing plenty of moisture while being established, it can survive dry spells. It is easier to grow from seedlings than from seed, or propagate it from an existing plant. Once a plant is well established, it will spread by sending out runners. Dig around the edge and cut off runners, making sure to use sections that have roots and foliage. Keep the roots moist until they are well established. If the plant gets too big, trim it and add the trimmings to the compost heap.

Growing in containers

Yarrow does not do well in containers.

Healing properties

Yarrow's Latin name comes from the mythological superhero Achilles. He had been a student of the centaur Chiron, who had great knowledge of medicinal uses of herbs. When he led his men into battle during the bloody Trojan War, Achilles remembered what he had learned and used yarrow to staunch the flow of blood from his wounded soldiers.

The white-flowered variety of yarrow is most often used in herbal medicine. Both leaves and flowers promote sweating and reduce fevers. They also stimulate digestion and are used to treat menstrual and urinary disorders. The leaves are an astringent and encourage blood clotting. The flowers contain powerful anti-allergenics making them a useful treatment for bronchial problems, hay fever and asthma.

How to use

Infusion: (flowers) To treat bronchial infections, irregular periods due to menopause, hay fever and asthma, rheumatism and arthritis.
Infusion: (flowers and leaves) To reduce fever and as a digestive tonic.
Tincture: (flowers and leaves) To regulate menstruation and treat urinary infections.
Steam inhalation: (fresh flowers) To ease chest congestion and asthma.
Compress: (infusion of flowers and leaves) To treat varicose veins.
Poultice: (leaves) To prevent bleeding on fresh wounds; in nostril to stop nosebleeds.

Culinary uses

Young yarrow leaves can be used in salads.

Cautions

Yarrow can cause skin irritation, especially after prolonged use. Do not take if pregnant.

THE HERBAL PHARMACY

Herbs for health

Many medicines commonly used today are the result of ancient knowledge. Aspirin is a good example. For millennia the bitter bark of the white willow tree was boiled and used to treat aches and pains and to reduce fever. In more recent times it was discovered that willow bark contains salacin, which nineteenth-century chemists synthesised into acetysalicylic acid, or aspirin.

The biggest difference between herbs and many modern medicines is that herbs help the body heal itself rather than just suppressing the symptoms. For example, many modern cold medicines treat just the symptoms by drying up the mucus. However, taking a herb such as echinacea helps strengthen the body's immune system, enabling us to fight off the cold more effectively and preventing us from being infected in the first place.

The other main difference between herbs and modern pharmaceuticals is that herbs are complex mixtures of many elements. Modern science prefers to isolate an individual effective element and synthesise it into a pharmaceutical, which can then be patented, branded and marketed. What is lost along the way is the beneficial result of the many elements mixed together – this combined effect is often more beneficial than any single element.

Many medicinal plants contain healing elements that have potentially dangerous side effects, but at the same time they contain substances that cancel out this possible danger. A good example is dandelion, which is a diuretic. Using a diuretic can result in a loss of potassium and – hey presto – dandelion leaf is rich in potassium.

Pounded into pastes, chewed, brewed, added to food and drink, and made into oils, gargles and lotions, herbs have been used by humans since we were dodging dinosaurs. Information about their use has been passed by word of mouth from generation to generation, the knowledge being continually refined and expanded

Many herbs dry well, retaining their flavour and efficacy. When using herbs you have grown and dried yourself, you can be assured that they are 100% organic without any pesticides

Herbs are inexpensive and easily available – especially if you grow them yourself. If you don't grow your own, you can buy herbs fresh from your local greengrocer or dried at most health shops. All herbal remedies work better if they are part and parcel of a healthy lifestyle, with good eating habits. In much the same way as for growing vegetables and herbs, where it all starts with creating healthy soil, so we need to start by building a strong foundation of a healthy body. Becoming more conscious of what we put in our bodies is the first step towards changing habits that could be the cause of an ailment or illness. Eating a variety of organic vegetables (with as much as possible being home grown), and making meat the side dish instead of the main (if you are an omnivore), both go a long way towards becoming healthier. If you do eat meat, become a more conscious carnivore by sourcing locally farmed, free-range chickens and grass-fed cattle. We have lost touch with what we eat and where it comes from. When last did you eat a meal where you could identify the origin of every single ingredient? This disconnect affects us more than we realise. Over the years, as I have become more aware of what I am eating and the more I use herbs for healing, the more resistant I have become to colds, coughs and other contagious bugs. Although I am a strong believer in herbal medicine, I also believe there is a place for modern pharmaceuticals. But as with everything, it is about making informed choices and doing things in moderation.

Becoming aware of what we put on our bodies is a further step towards creating a healthy foundation. Our skins are permeable and they absorb what we put on them. Harsh chemicals and preservatives contain substances that can be harmful to us. Many of these elements remain in our bodies, stored in our liver and body tissues, affecting us for years after we used them. Start informing yourself by reading labels and understanding more about the ingredients in the products you use.

Many people, when I suggest they should try using herbs, respond by saying it is too much of a hassle to even think about taking time out of their busy days to chop and mix a herbal tea. Popping a pill is a much quicker solution. However, it does not take long to mix up a batch of herbal tea in the morning and put it in flask so you can drink three or four cups a day without any bother. In the recipe section (page 198) there are plenty of other recipes for using herbs on the go. And bear in mind, if you can't take the time out to make a simple cup of healing tea, that in itself is probably the cause of many an ailment.

Using herbs with intent

People often ask me if I talk to my plants. In reply I say, 'not often, but I listen to them all the time'. I have spent so many hours in my garden observing the smallest details and absorbing the energy that I have come to know my plants well.

Many people who live close to nature develop an intimate knowledge of plants and plant lore. One of the richest and most diverse sources of healing plants is the Amazon forest, where hundreds of plants are used as medicine, and shamans are living encyclopedias of knowledge on how to prepare plants to treat a wide range of ailments. What I find wondrous is that the plants themselves are the teachers. Shamanic healers take a combination of plants to go into a trance. While in this state, the shaman goes on a 'journey' into the forest, where he is shown a plant he might never have seen or used before. He will be told how to prepare and use the plant. Once out of the trance, the shaman walks into the forest, following the path he trod while in a trance, to the exact plant he was shown. And, by following the instructions, he heals his patient.

Here in South Africa, acacia trees are targeted by giraffes for their tasty leaves. However, the acacia objects to being eaten and as soon as it feels a giraffe beginning to munch its foliage, it produces tannins to alter the flavour from sweet to bitter. And, as a favour to its neighbouring acacias, it sends pheromones out into the air saying: 'Giraffe alert! Giraffe alert!' The nearby acacias pick up these messages and they quickly change the flavour of their leaves. Plants do have the ability to communicate – we just have to learn how to listen.

While I was writing this, I met a young man when I was out walking my dogs in the park. His name was Bruce and he had a badly injured leg from a bike accident. I asked him if he was using any herbal remedies and he said he was using both comfrey and calendula. We began chatting about herbs and he mentioned yarrow. Just the day before I had been reading in Maria Treben's book *Health from God's Garden* about how yarrow stimulates growth in bone marrow. As I began to mention this to Bruce, he finished my sentence saying, 'Yes, yarrow stimulates red blood cells in bone marrow.'

'Oh,' I said, 'I only just read about that yesterday.'

He looked at me a little oddly and replied, 'So did I.'

I started smiling and took a chance, 'You must be reading Maria Treben's book then?'

'Yes,' he said, '*Health from God's Garden*.'

Coincidence? I think not.

When making and using herbal remedies, intention and state of mind are potent factors that can influence the efficacy of healing, and should not be underestimated. The placebo effect has been well documented: patients are given an inert pill, but because they believe it contains medicine they see a marked improvement in their condition. This shows the powerful role our minds play on our physical wellbeing. And it works both ways – if you are trying to heal someone who is very sceptical about the herb's ability to heal, chances are the remedy won't work as well. A pharmacist friend would say to me, 'If you take arnica, the bruising will go in two weeks. If you don't take it, the bruising will be gone in a fortnight.' And his bruises usually lasted about two weeks. On the contrary, if we respect the power of the herb, and believe in its ability, it seems to open our bodies to its healing powers, making them stronger and more effective.

Yarrow planted in a medicinal herb garden will increase the efficacy of neighbouring herbs

Medicinal herbs and safety

As with any medicine, herbs should be treated with respect. Just because they are 'natural' doesn't mean they aren't harmful – the very reason they work is because they contain powerful constituents. Please read the information on each specific herb before taking it. For example, some herbs should not be taken by pregnant or breast-feeding women. It is also not recommended to treat children under the age of two years. Homemade herbal remedies can safely be used to treat everyday ailments such as colds, cuts, stress, bruises and burns. Herbs are also a good option if you have a chronic condition that is not responding to pharmaceuticals. For more serious conditions, consult a trained herbalist or a doctor.

Some herbs can cause an allergic reaction. If you are an allergy-prone person, do a skin test before using the herb. Take a dab of the herbal product and rub it onto a small section of skin on your inner arm. Wait for 24 hours to see if a rash, redness or itching occurs. When taking a herb internally, start with small dosages and build up to the full dose. If you develop itchiness, a rash, an upset tummy or headaches when you use a herb, stop using it immediately.

Herbs can also react with other medication. If you have an existing medical condition or you are taking any medication, consult a trained herbalist before starting to self-medicate with herbs. Lastly – and most importantly – make sure that the herbs you are using have been grown organically with no pesticides sprayed on them and that you have identified them correctly.

Diagnostics and dosage

In this age of convenience and specialisation we expect instant solutions. We go to the doctor with a sore stomach expecting to be diagnosed, treated and made better. But it doesn't always work this way. The doctor cannot know everything about your body. However, if you listen to your own body, you can get to know it very well. As you become more attuned to it, you will recognise if a sore stomach is because of something you have recently eaten or if it's due to stress.

When using herbs to medicate, the first step is to diagnose the problem. This involves observing and listening to your body and the symptoms of its ailments. If you have a cold, is it predominantly a runny nose or is it more of a sore throat? That rash on your leg – have you recently brushed against something or does it occur when you are stressed? Armed with this knowledge, you can then source the best herb to achieve the optimum effect. But which one to choose? Many herbs have more than one application and we all have different reactions to them. So each herbal solution will be based on the individual and their specific needs. And it is not that difficult.

Most of us will be treating a small circle of people and for that you don't need to learn about hundreds of symptoms and herbs – you just need to get to know a few that work well for you, your family and friends. Don't place pressure on yourself to learn and remember all this information – this is why we have reference books! Long before I ever thought about writing any books, I bought a small A–Z notebook and began keeping track of what herbs I had grown and what ailments I used them for. A notebook is very useful to keep records of recipes, herbs you have used and how well they worked. I'm not a trained herbalist. However, I enjoy growing and using herbs, and I have found the more often I use a herb, the more familiar it becomes.

It is best to start treatment with one herb at a time so you learn the effects of each one. This will also minimise the chance of allergic reactions. Using herbs is not an exact science and is open to experimentation. Herbs are administered in many different ways, from teas to tinctures, ointments to decoctions. In 'Herbal preparations' (page 188) I have provided detailed instructions on how to make the various types of preparation.

Herbs for beauty, home and garden

Herbs have a multitude of uses other than healing. As with any modern pharmacy, the shelves of a home herbal pharmacy should be stocked with cosmetics, pet preparations, household products and garden remedies. This is where the fun and pleasure of herbs really come into their own. Homemade products can be far more luxurious than bought ones. As we are not paying extra for the packaging, advertising, transport and mark-up, we can afford to splurge on ingredients and use the resulting delicious product liberally.

I have been making my own lotions and potions for years now, because it is fun and rewarding. Although self-sufficiency is an appealing idea, for most of us it is unrealistic. However, it is very empowering to have the knowledge to make my own body cream or hair conditioner. The recipes in this book (starting on page 198) are all easy and fun to make, and are a starting point for you to begin experimenting with your own mixtures.

Herbs in the kitchen

And where would we be without herbs in our kitchens? Even a small selection of home-grown herbs can make a huge difference to a meal. Simple grilled chicken breasts become something special when they are topped with a tarragon and lemon butter sauce and a tomato salad is elevated to new heights with the addition of basil chiffonade. Even if you use herbs for culinary purposes only, it is worth growing a selection. Bought herbs never last for long and they are so expensive – especially when you realise just how easy most of them are to grow.

Stocking the Herbal Pharmacy

BASIC INGREDIENTS

Basic cream

A basic cream is useful when making lotions and body creams, but ensure the one you choose is vegetable based. Preferably don't use aqueous cream as this is made from petroleum by-products and contains the detergent sodium lauryl sulphate, which is detrimental to our bodies. Although it was originally developed as an alternative to soap, aqueous cream has become commonly used as a moisturiser. Research has shown that it actually exacerbates skin conditions such as eczema. Regular use of aqueous cream on healthy skin reduces its thickness and dries the surface out. Rather look for a vegetable-based cream in health shops, such as the one made by W-Last Homeopathic.
Available from health shops.

Beeswax

Beeswax was one of the earliest products used by humans for a multitude of purposes. It was used as a sealing and waterproofing agent for houses and boats; to make candles for lighting; for coating wooden writing tablets; in cosmetics and herbal preparations; and as a modelling agent for everything from sculpture and jewellery moulds to dentistry.

Beeswax is what bees use to make honeycomb. A by-product of harvesting honey, it is a healthy, natural moisturiser, used in body lotions, lip balms and face creams. It works well in skin products because it contains compounds called wax esters, which also exist in human skin. Plus it has healing and antiseptic qualities.

In addition to its health benefits, when mixed with borax, it is an emulsifying agent used to mix oils and water together into a smooth cream. Its melting point is 62–65°C. It should be melted using a double boiler, or a bowl set over a pan of simmering water, as it is flammable if it comes into contact with direct flames. It is ideal for making candles as it is very slow-burning and gives off a lovely honey scent.
Available from health shops and from beekeeping associations.

Borax

This naturally occurring alkaline mineral has been used for thousands of years for a multitude of purposes – from an additive in pottery glazes to a curing agent for caviar. It is also an effective emulsifier when combined with beeswax. This is particularly useful to emulsify body creams that contain both oil and water. Work with caution when handling borax, as the raw powder can irritate the skin or respiratory system if inhaled.
Sold as a powder in the cleaning section in supermarkets.

ABOVE: A by-product of harvesting honey, beeswax is a healthy, natural moisturiser, used in body lotions, lip balms and face creams.

ABOVE: Cocoa butter is the fat extracted from cocoa beans. It has been used for centuries as an effective moisturiser.

BELOW: Honey is a beneficial addition to bitter herbal teas, sweetening them and adding to their healing power.

Cocoa butter
This pleasant-smelling butter is the fat extracted from cocoa beans and has been used for centuries as a moisturiser. It melts at body temperature and is easily absorbed by the skin, keeping it soft and supple. It is particularly good for itchy skin conditions such as psoriasis and eczema. *Pure cocoa butter is available from health shops.*

Coconut oil
This is extracted from the white flesh of the coconut. In cooler conditions it solidifies into a thicker butter. It has a very small molecular structure, meaning it is easily absorbed by our bodies. It is used extensively in India for hair care, as it is an excellent hair conditioner, increasing shine and helping damaged hair to grow. It is also an effective moisturising massage and body lotion, helping with the treatment of various skin ailments such as eczema and psoriasis. It helps prevent premature aging by reducing wrinkles and tightening sagging skin. *Available from health shops and Indian grocers.*

Gelatine
This thickening and setting agent is used to make gels. *Available from supermarkets.*

Glycerine
This viscous, odourless liquid is used to thicken and preserve creams and lotions. It is sweet tasting and has soothing properties, useful for throat and digestive ailments. *Available from pharmacies and health shops.*

Gum Arabic
This edible resin, from a North African acacia tree, has glue-like properties that bind liquids together to create chewable pills or lozenges. Acacia gum soothes irritated and inflamed mucous membranes. Added to lozenges, liquids or ointments, it helps heal wounds, ease sore throats and settle gastric upsets.
Available from Indian spice shops and health shops.

Honey
Honey's therapeutic properties have been known for thousands of years. Externally its anti-inflammatory and antimicrobial abilities help heal wounds, burns and fungal conditions. Internally its cleansing and antioxidant properties heal colds, flu and throat and bronchial conditions. It is particularly effective in treating pollen-related allergies – especially if treated with locally produced honey. My bad spring hay fever was vastly reduced after I began keeping bees and eating a daily teaspoon of honey produced from my own garden. Honey is a beneficial addition to bitter herbal teas, not only sweetening them, but also adding to their healing power.

Oils

Oils are used either as a base to infuse herbs or are added to body lotions and other products. The base oil depends on the end product. For creams and lotions, use non-greasy, neutral oils such as sunflower, grape seed or safflower oil. If you are making lip balm or body cream for very dry skin, olive, wheat germ or avocado oil will be a better choice. Other oils that can be used are jojoba, evening primrose, vitamin E, walnut, coconut and almond. For edible infused oils, use good-quality olive, walnut, macadamia, grape, avocado or sunflower oil.

Shea butter

This wonderful moisturiser is extracted from karité trees (*Vitellaria paradoxa*) in Central and West Africa. Unrefined shea butter has healing and moisturising properties. It helps our skin absorb moisture from the air and has antifungal, anti-inflammatory and sun-screening properties. It enhances body lotions, hair conditioners, shampoos and soaps. It helps treat skin allergies, eczema, psoriasis, fungal infections, dermatitis and itchy, dry skin and reduces scarring, stretch marks and wrinkles.
Unrefined shea butter is available from health shops or online.

Soap

Soap is a surfactant, which means that it mixes with both oil and water. Most dirt is attracted to oil, such as our skin with its oil layer. If we wash with just water, it won't remove all the oily dirt. By adding soap it penetrates the water and separates the dirty oil molecules so they can be washed away. Originally soap was made using natural products (fat and leached wood ash). Recipes for soap have been discovered on Egyptian papyrus dating back to 1500 BC and soapy residue has been found in Babylonian clay cylinders from 2800 BC. Today, however, what we know as soap is most often not the same thing the Ancient Egyptians used. Detergents made from petroleum products and chemicals have almost completely replaced natural soap (even though they are still labelled soap). Because these chemicals have a nasty odour, synthetic fragrances are added. They also contain preservatives and antibacterial ingredients. No wonder, then, that many people suffer from skin allergies when using them. When using a soap base to make soap at home, look for a pure, natural soap made with oils such as coconut or olive.
Available from specialist soap shops or from health shops.

Witch hazel

An extract from a low-growing shrub native to North America, witch hazel is an astringent often added to herbal face toners and aftershaves. It draws out toxins, which helps treat pimples, acne, insect bites and stings. It is an anti-inflammatory and it helps shrink blood vessels back to their normal size, effectively treating haemorrhoids.
Available from pharmacies and health shops.

PRESERVATIVES

Many commercially prepared body lotions contain preservatives and chemicals to prevent them going rancid or spoiling. This means they can keep for years. But these preservatives and chemicals can have many unwanted side effects, which is why I choose to make my own chemical-free lotions. Without the preservatives, homemade lotions can go off quite quickly, resulting in an unpleasant (and sometimes unhealthy) product. Here are a few tips to reduce the chances of this:

- Make sure your equipment and containers are sterilised (see page 188).
- Store the finished product in airtight dark or opaque glass jars.
- Make small batches.
- Keep your ingredients and raw supplies in a cool dark place. Store any extra in the fridge until you are ready to use them.
- Products made without water last longer.
- If you are adding water, preferably use distilled water, and include a natural preservative (see list below).
- Wash your hands before using products.

With natural products, decay is inevitable. There are no natural preservatives that will completely halt this. However there are some natural products that will fight bacteria and mould, slowing spoilage

These natural products will fight bacteria and mould, and slow spoilage. They are all available at health shops, pharmacies or online.

Aloe vera Antimicrobial and antifungal.
Avocado oil Antioxidant.
Borax An emulsifier, preservative and cleanser, borax is often used in shampoos and bath products. When combined with citric acid in bath salts or bombs, it makes them fizzy.
Cinnamon Inhibits bacteria and slows growth of mould, yeast and fungi. Use pure powder rather than essential oil as oil can irritate skin.
Citric acid/vitamin C powder Powerful antioxidant and natural preservative that prevents rancidity and the growth of bacteria. It can also be used to adjust the pH of a lotion so it matches our body's natural pH of 6 to 6.5.
Geranium essential oil (Extracted from rose-scented pelargonium) Antioxidant and antibacterial. Also slows growth of moulds, fungi and yeast.
Goldenseal root Inhibits growth of moulds, fungi and yeast.
Green tea extract and powder A very strong antioxidant.
Honey Antioxidant and antibacterial.
Jojoba oil Antibacterial and antimicrobial.
Lemon grass essential oil Antiseptic, antibacterial and antifungal.
Olive leaf extract Antifungal and antimicrobial.
Tea tree oil Antibacterial and inhibits growth of moulds, fungi and yeast.
Thyme essential oil Antimicrobial and inhibits growth of moulds and fungi.
Vitamin E oil Powerful antioxidant.

TOOLS AND EQUIPMENT

You don't need to go shopping to make most of the herbal remedies in this book – you will probably already have everything you need. Here is a list of things you will find useful:

Blender or processor To chop up herbs finely.
Bottles and jars Instead of buying new bottles and jars, start recycling used ones. Just make sure they are washed well and sterilised before use. If you are reusing a bottle that had a cork, rather replace the cork with a new one. Don't forget to label your bottles.
Cheesecloth, muslin or paper oil filters For filtering herbs from various concoctions.
Double boiler Many of these preparations are made using ingredients that should be melted or heated very slowly and gently, and a double boiler is the best tool for this. It has two levels: one below (containing water) that is in direct contact with the heat, and one above, where the ingredients are placed and are slowly heated by the steam from below. If you don't have a double boiler, use a saucepan and place a glass bowl over the top, making sure the glass bowl does not touch the water in the pan below.
Funnels and colanders Various sizes are useful to decant and strain herbs.
Grinding mill Useful for grinding up hard dried herbs and spices – a recycled spice bottle with a grinder top will do.
Notebook Essential tool to keep track of what you are doing.
Pestle and mortar To pound herbs into a paste.
Spatulas and scrapers To decant ointments and balms into their containers.
Tea infusers and strainers To quickly make herbal infusions. There are various varieties available: single-serving ones that you fill and place in a cup of boiling water, cups with a removable strainer top section (often available at Chinese supermarkets as they are used for loose green tea), or teapots with a central strainer section for larger quantities.

SOURCING HERBS

If you don't have the required herb growing in your garden, there are many specialist herb and health shops where you can find everything you need. Most of the people working at these shops are very knowledgeable about herbs. There are also numerous online sources for herbs.

If you don't grow your own herbs you can buy them fresh from your local greengrocer or dried at most health shops

THE HERBAL PHARMACY

Herbal preparations

BEFORE YOU START

Because homemade herbal preparations don't have strong preservatives protecting them against bacteria, microbes and fungi, the first step to prevent these is to sterilise the bottles. Wash the bottles and lids in hot soapy water and rinse well. If you are using bottles with screw-top lids, place them upside down on an oven-proof tray, so any water drains out. Place the lids alongside the bottles on the tray. Bake at 160°C for at least 10 minutes. Remove and leave to cool before filling.

An alternative method to baking the bottles in the oven is to place the washed, wet bottles upside down in a microwave on high for 3 minutes. If you are using bottles with rubber rings or metal lids, which can't be heated in an oven or microwave, place the bottles, lids and rings on a rack in a deep pot. Cover them with water, bring it to the boil and simmer for 10 minutes. Remove them with tongs and place them on a tray covered with a clean dishcloth to drip dry. Be careful when handling hot jars – they can hurt you! You also need to thoroughly clean all your equipment before using it, by boiling it in hot water for 10 minutes.

When you start playing around making infusions, creams and ointments, it is a good idea to keep a record of what you are doing. In a year's time you will never remember what you added to that great-smelling hand cream that everybody loved. In your journal, write down information about each batch you make. You can simply include the date and ingredients or, to keep more accurate records, you can include some of the following: herb used (and which part of the herb was used), date of harvest or origin of herb (if not from your own harvest), dried or fresh, date it was dried, date it was made and quantities.

ABOVE AND BELOW: Making balms and body creams is fun and rewarding.

Creams and lotions

A basic cream (or lotion) contains water mixed with fats or oils. As we know, these two don't mix, so you must use an emulsifying agent. Preferably avoid using emulsifying ointment, which consists of emulsifying wax, soft paraffin and paraffin oil. Paraffin is a by-product of petroleum and because of its large molecular structure it blocks essential oils and nutrients from entering the skin. It dries the skin and also blocks waste from exiting pores. Emulsifying ointment is the main ingredient in aqueous cream, which also should preferably not be used for natural herbal preparations.

The best natural emulsifier is beeswax mixed with borax. Only a small amount of borax is needed to create the chemical reaction that emulsifies the beeswax in water. This results in a cream that is easily absorbed by the skin, without feeling greasy. The proportions can be tricky to get right. If you use too much borax, the mixture will become gritty, but if you add too little, it doesn't emulsify. When starting out making body creams, rather try making balms first, as they don't contain water and don't need an emulsifier.

Basic body cream

First melt the beeswax in the top of a double boiler or glass bowl set over a pan of simmering water. Once the wax is melted, add any oils, other than essential oils, and stir to mix. In a separate bowl, mix the water with the other liquid ingredients such as herb extracts, aloe vera, basic cream, witch hazel or glycerine. Add the borax and heat until it is dissolved and the mixture is hot but not boiling (about 75 °C). The next stage is very similar to making mayonnaise. Add the melted beeswax and oil (it helps to transfer it to a jug first) to the hot liquid in a very slow, steady drizzle, stirring constantly as it thickens. If the mixture does separate while cooling, add a quarter cup of basic cream at a time, whisking it in, until it is smooth again.

Decoctions

Decoctions are most often used for more robust plant parts: roots, seeds, stems, bark and berries. These hardier parts of the plant need to be simmered to extract the beneficial constituents. When using freshly harvested roots, cut off any aerial parts and wash the roots well before chopping them. When using seeds, crush them first. The following recipe is sufficient for three doses. The standard dose is 1 cup, three times a day, drunk hot or cold.

Basic decoction

- 30 g dried herb or 60 g fresh herb • 750 ml water

Chop the herb and add to cold water. Heat and simmer gently for 1 hour (it should be reduced by about one third) and strain. This should be made fresh daily.

ABOVE: Dried angelica root is used to make a decoction for alleviating menopausal symptoms.

Dried herbs

For instructions on how to dry freshly harvested herbs, see page 36. When you buy dried herbs from a health store, decant them into an airtight, glass bottle and store in a cool spot.

Essential oils

An essential oil is a liquid distillation of a plant. Essential oils are very concentrated and can irritate the skin and should be diluted with a carrier oil such as almond or wheat germ oil before applying. For essential oils in a bath, add 5 to 10 drops of essential oil to 20 ml of carrier oil and mix. For a massage oil, mix 10 to 12 drops with 30 ml of carrier oil. To add to an oil burner, start with 5 to 10 drops and add as needed. Most essential oils should not be taken internally.

Herbal honeys

Herbal infused honey is a versatile way of preserving and using herbs. Because it has antibacterial properties and boosts the immune system, honey increases the medicinal effect of the herb. Herbal honey can be taken internally (a great way to give children a bitter-tasting herb) and it can also be used externally as a poultice to treat burns and infected wounds. Because honey is soothing, it is a good choice to mix with herbs that are being used to treat the throat and respiratory system. Use it as the base for a cough syrup or to make cough sweets.

Making herbal honey
Start with raw honey as it has the most health benefits. If you are using fresh herbs make sure they are as dry as possible to avoid mould forming. (I have never had a problem with mould on my herbal honey when using fresh herbs but in a more humid climate it might be an idea to keep herbal honey in the fridge.) Fill a sterilised bottle with roughly chopped dried or fresh herb. If you are using a woody herb or stems and roots, chop it a bit more finely. Cover with honey, poking with a sterilised spoon to remove any air bubbles, and seal tightly. Leave it to infuse for about six weeks. Strain and reseal. (You might need to warm the honey slightly to pour it but don't heat it too much.)

Good herbs and spices to use for herbal honey are: cayenne pepper, cinnamon sticks, echinacea (root), elderberry (use cooked berries and mush them before infusing with the honey), fresh ginger root, garlic, lavender, liquorice, lemon balm, lemon grass, lemon verbena, marjoram, onion, oregano, peppermint, rose-scented pelargonium, rose petal, sage, thyme and turmeric (try to get hold of turmeric root rather than powdered).

Infusions
This is one of the simplest ways to use herbs, as many of the valuable constituents are soluble in water. An infusion is made in the same way as you make tea – infused in just-boiled water. Don't add vigorously boiling water to herbs as valuable oils will be lost in the steam. This method is best used for leaves, flowers and soft stems. Stevia or liquorice root (available at health shops), sweeten bitter infusions. Liquorice adds many health benefits, but don't drink excessively if you have high blood pressure. Infusions should be made fresh every day. The standard dose is 1 cup, three times a day, drunk hot or cold.

Standard infusion
- 30 g dried herbs or 75 g fresh herbs • 500 ml just-boiled water

Put the herb in a teapot. Add the water and leave to infuse for 10 minutes. Strain. You can also make a cup as you need it, using an immersible strainer or teacup with built-in strainer (see page 186).

Juicing
Juicing is an option only if you have a really large harvest: a 10-litre bucket full of herbs will produce only about 100 ml of juice. Blend the herb in a processor until it is pulped. Dollop into a piece of cloth and twist it into a ball. Squeeze the juice out through the

Infusions (opposite) are best used for leaves, flowers and soft stemmed herbs, such as lemon verbena (above).

cloth. The pulp can be used in other herbal preparations, as it will still have plenty of goodness remaining.

Oil infusions

Oil infusions are made by steeping the plant material in oil to extract the beneficial elements. They are easy to make at home and don't need the specialised equipment required for essential oil extraction. Infused oils are similar to essential oils but not as concentrated and don't have to be as diluted when using. They can be used on their own as massage or culinary oil (using edible herbs) or they can be used as a base for creams, balms, ointments or bath oils.

Infused oils will last up to a year if they are kept in a cool, dark place. However, the potency will be reduced and, depending on the oil used, they can go rancid. It is better to make smaller amounts more often and keep them in the fridge during warmer weather. When making products using infused oil, their shelf life can be increased if a natural preservative is added (see page 185). As with all herbal preparations, make sure all your equipment and storage bottles are sterilised. You can use a variety of oils, from sunflower or olive oil to jojoba or almond oil. Some oils have medicinal properties of their own that will add to the healing properties of the infused herb. The choice will depend on the end use: culinary herbs, such as basil for use in salad dressings, can be infused in good-quality olive oil. For hair conditioner with rosemary-infused oil, jojoba will be a better choice. Other oils you can use are coconut oil, grape seed oil and flax seed oil. Using herb-infused oils is an easy way to add instant flavour to a meal. It is also a good way to preserve a summer herb into winter. (For more on oil see page 184.)

Chop herbs up as finely as possible before infusing. Dried or fresh herbs can be used and there are advantages and disadvantages to both. Using dried herbs helps speed up the infusion process, as the herb's membranes are already collapsed and the oil can get in and start dissolving the oil-soluble elements more quickly. Another advantage of dried herbs is their reduced water content, which lessens the chances of moulds and fungi developing. However, fresh herbs do contain more elements than dried herbs, as some of the more volatile ones are lost in the drying process.

There are two methods of oil infusion: hot and cold. A hot oil infusion will be able to be used more quickly but a cold oil infusion will retain the medicinal qualities for longer.

Hot oil infusion

For a hot oil infusion, use dried or fresh herbs in roughly the following proportions:
- Fresh: 1.5 parts herb to 2 parts oil
- Dried: 1 part herb to 2 parts oil

Mix the oil and herbs together in a double boiler or a bowl set over water and heat it gently for about 3 hours. Strain the mixture into a bowl or jug. Use a funnel to pour it into a sterilised, airtight glass bottle.

ABOVE: Oil infusions are similar to essential oils but are not as concentrated and they don't have to be as diluted when using.

Cold oil infusion

Pack a sterilised bottle with dried or fresh herbs and cover them completely with oil. Seal and leave in a sunny spot for 2 to 3 weeks. Strain the mixture into a bowl or jug, squeezing to get most of the oil out. Either store the strained oil or, for a double-strength infusion, pack the bottle a second time with the same type of herb, but this time cover with the infused oil.

Seal and leave to steep in the sun for 2 to 3 weeks again. Strain and bottle it. For a triple-strength infusion repeat the process again before bottling it.

Ointments and balms

Ointments or balms are made in a same way as creams or lotions but without the added water. They have a thicker consistency and are used for many skin treatments.

Poultices and compresses

Both of these are applied directly to the skin to speed up healing of wounds; to treat skin conditions, headaches, varicose veins, bruises, sprains, muscle aches and joint pains; and to draw out pus and toxins from abscesses or boils. The difference between the two is that for a compress, a cloth is soaked in an infusion or decoction of the herb. For a poultice, the actual herbs are used, held in place with a bandage or cloth.

Compress

Soak a cloth in a hot infusion or decoction. Wring it out and hold the cloth over the affected area until it cools. Repeat as often as required for 10 to 15 minutes at a time. Do not use for longer than two days.

Fresh herb poultice

Place ½ cup of herbs in a saucepan and add 1 cup of boiling water. Simmer for 2 minutes and remove from the heat. Place a piece of cloth on a flat surface. Scoop out the herbs, squeezing out any surplus water, and place them on the cloth, wrapping it over the herbs to make a parcel. Place this on the affected area and secure it with a bandage or safety pins and leave for 2 to 3 hours. Repeat 2 to 3 times a day for no longer than three days.

Dried herb poultice

Grind the herbs in a pestle and mortar until they are powdered. Place them in a bowl and add just enough hot water to form a paste. Apply in the same way as for the fresh herb poultice.

Sprays

Herbal sprays can be made in a number of ways. It is best to keep a herbal spray in the fridge to prevent spoilage.

ABOVE: Herbal balms are used for many skin treatments.

Essential oil spray
Mix 20 drops of essential oil with medicinal alcohol. Add ½ cup of water. Decant into a spray bottle. Experiment with different combinations of oils. This recipe is not to be used on the body or internally – it is used to spray surfaces and as air freshener.

Fresh or dried herb spray
Chop up 1 cup of fresh herbs or ½ cup dried herbs and place in a bowl. Add 2 cups of boiling water and leave to infuse overnight. Strain and dilute further with another 2 cups of water. If you are making a spray to use in the garden, add a small amount of dish-washing liquid to help it stick to the leaves.

Steam inhalation
Mixing fresh herbs, infused oil or essential oil with boiling water and inhaling the steam will ease congestion in sinuses and lungs. It is also a beneficial facial, opening and deep-cleaning the pores. Repeat the process two to three times a day. Chop the recommended herb and place it in a heat-resistant bowl. (I use a metal bowl placed inside another bowl as this helps it stay hotter for longer.) Add the boiling water. If you are using infused oil or essential oil, add it to the water. Cover your head with a towel, making sure the sides of the bowl are covered so no steam escapes. Breathe slowly and deeply. If your sinuses are very congested, try holding one nostril closed and breathing through the other, alternating nostrils. Or breathe in through your mouth and slowly out through alternate nostrils. Have tissues on hand! For chesty conditions, breathe deeply through your mouth, filling your lungs with steam. Keep yourself warm when you have finished.

Apple cider vinegar mixed with water can also be used for steam inhalation, either plain or with herbs added.

ABOVE: A steam inhalation using tea tree oil will ease congestion in sinuses and lungs.

Syrups and sweets
Honey or sugar can be used to preserve a decoction or infusion. The sweetness also helps to hide the taste of unpleasant-tasting herbs. Syrups can be made with herbal infusions or decoctions, or they can be prepared using fresh herbs. Sweets are generally made using infusions or decoctions. For the syrup, take 1 to 2 teaspoons three times a day. Suck a sweet three times a day.

Syrup using a decoction or infusion
- 500 ml decoction or infusion
- 500 g honey or sugar

Heat the decoction or infusion in a pan and add the honey or sugar. Stir until mixed or the sugar is dissolved. Heat until it just starts simmering. Remove from the heat and leave to cool. Pour into a dark glass bottle and seal with a cork. (Syrup can ferment and if it does, a screw-top bottle will explode.)

Syrup using fresh herbs
- 1 litre water - 500 g honey or sugar - 150 g herb

Heat the water in a pan and add the honey or sugar. Stir until mixed or the sugar is dissolved. Add the herbs and heat gently until it just starts simmering. Remove from the heat and leave to steep overnight. Bring to a simmer again. Remove from the heat, strain and pour into a dark glass bottle and seal with a cork. (Syrup can ferment and if it does, a screw-top bottle will explode.)

Sweets
- 200 ml infusion or decoction - 200 g sugar - 2 drops edible essential oil

Mix the infusion or decoction and sugar in a heavy pan over low heat, stirring until the sugar is dissolved. Turn the heat up and boil without stirring until it cracks when a small piece is dropped in cold water (about half an hour). Add the essential oil and pour into paper baking cases or sweet moulds. Leave to set. Remove from moulds, toss with icing sugar and store in an airtight container.

Herb honey balls
- honey - fresh or dried herbs, chopped finely in a blender

Add the herbs to the honey, stirring until it forms a very thick paste. Pinch off small pieces and roll them into balls. Roll in icing sugar and leave to dry on a tray for a day or two before sealing in a container.

Tinctures

A tincture is more potent than a decoction, infusion or dried herb and it lasts longer. Alcohol is used to extract the herb's active ingredients and preserve them for up to two years. Vodka is an ideal tasteless alcohol. Rum or brandy are good for disguising the taste of bitter herbs. Do not use rubbing alcohol – it is poisonous.

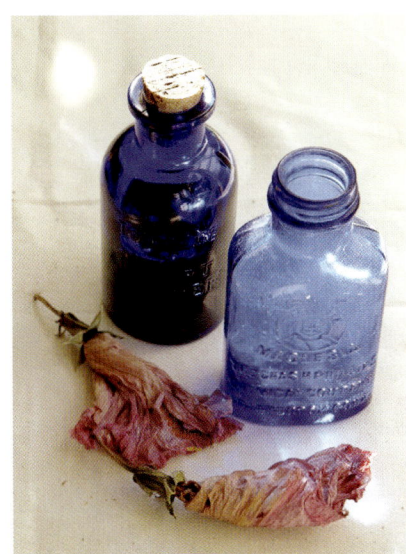

Standard tincture
- 200 g dried herb or 600 g fresh herb, chopped - 1 litre 80 to 100 proof alcohol

Place the herb in a large bottle and cover with the alcohol. Seal and leave in a cool place for 2 weeks. Give it a shake every few days. Strain into a bowl. Wrap the herb in cheesecloth and twist it tightly to squeeze the last of the tincture out. Use a funnel to pour into a dark glass bottle and seal.

The standard dose is 5 ml, three times a day. Tinctures should be taken diluted in water. Improve the flavour by adding honey or fruit juice.

Vinegars

Vinegar has been around for as long as wine. In fact, its name comes from the French *vin aigre*, which means sour wine. Versatile vinegar can be used in a multitude of ways: as a delicious condiment, a preservative, a medicine or as a household cleaner and disinfectant.

Herbal infused vinegars are easy to make and the possibilities of flavours and herbal combinations are endless. They should be stored in a cool, dark place as a warm environment can cause fermentation. Herbal vinegars will keep for up to two years.
Note: Don't use bottles with metal lids – vinegar will corrode them.

Culinary herb vinegars

In the kitchen, vinegar can be infused with herbs to create a wide range of flavours. The tastiest vinegars will result from using the best ingredients. White or red wine vinegar, malt vinegar or apple cider vinegar are all good choices. Use herbal vinegar to zingle a salad dressing, to perk up a marinade or to add a twist to pickled vegetables.

To make: Use fresh herbs, but spread them out in a warm place to dry for a day to reduce their moisture content. Place ½ cup of partially dried, finely chopped herbs in a sterilised bottle; pour 2 cups of vinegar over the herbs. Seal and leave to infuse in a dark spot (a cupboard for example) for at least 3 weeks, shaking every now and then. Strain the vinegar and decant into a storage bottle.

Medicinal herb vinegars

Herb-infused vinegar made for medicinal purposes is more concentrated than culinary vinegar. Apple cider vinegar is the most useful medicinally. Taken on its own it has powerful medicinal qualities – it is antimicrobial, antiseptic, antifungal, antispasmodic, antibiotic and astringent. Inhaling the steam from apple cider vinegar will soothe bronchial spasms and clear sinuses and a blocked nose. Vinegar can soothe tired and aching bodies, treat urinary tract infections, take the pain away from a sore insect sting, relieve headaches and clear dandruff. It prevents infections, relieves sore throats, eases digestive upsets and soothes sunburn. When you add to this the medicinal value of the herbs, herbal vinegar packs a powerful punch.

To make: Use fresh herbs, but spread them out in a warm place to dry for a day to reduce their moisture content. Fill a sterilised bottle with the partially dried herbs. Add apple cider vinegar until it covers the herbs. Seal and leave to infuse in a dark spot (a cupboard for example) for at least 6 weeks, shaking every now and then. Strain the vinegar and decant into a storage bottle.

Other uses for vinegar

Vinegar has plenty of other uses around the home. It is a solvent, and therefore an excellent all-purpose cleanser and disinfectant. It will whiten whites if you add vinegar to your rinse water. It removes stains, cleans brass, copper and pewter, and removes odours from fridges or rooms.

HERBAL RECIPES
for wellbeing, healing and happiness

Over the years I have developed a number of herbal recipes for cooking, cleaning, healing and feeling good. Following are some recipes to get you started. When using herbs there are endless combinations of flavours, scents and healing remedies that you can try. If you don't have the herbs in the recipe, experiment using the ones you do have. There are many with similar qualities. Use the A–Z section starting on page 40 as well as the chart on page 260 to create recipes to suit your needs, drawing on the herbs you have available. And don't forget, this isn't a science exam – it is supposed to be enjoyable and fun.

I have been experimenting with herbs for about 15 years now. Every time I bring a new plant home it becomes the flavour of the month as I try out different ways of using it. Many herbs grow in such abundance that they are just asking to be put to good use in a pesto or hand cream, an edible oil or a healing infusion

TASTY TEMPTATIONS   200

AROUND THE HOUSE   212

DELICIOUS BEAUTY CARE   216

COMMON AILMENTS
   Coughs, colds, throats, flu and hayfever   230
   Aches, pains, sprains and bruises   238
   Digestion, the stomach and urinary system   242
   Tension, stress, mind and emotion   245
   Heart, blood and circulation   247
   Women   249
   Skin conditions   250

JUST FOR PETS   254

IN THE GARDEN   258

Tasty Temptations *Tasty, zesty, delicious and healthy*

Herbs and spices not only add great flavour to our food, they have many health benefits too. They perk up bland dishes and are packed with vitamins and minerals. Using herbs and spices expands the variety and flavour of what we eat and reduces the amount of salt and sugar we use. The effects of regularly eating herbs are wide ranging: from building a healthy immune system and improving our digestive systems, to increasing energy levels and our feeling of wellbeing.

ELDERFLOWER VINEGAR

This healthy vinegar embodies the fresh taste of summer and is a good way of preserving seasonal elderflowers. Use it in salad dressings or mixed with honey and lemon for a hot toddy to treat colds and flu.

- elderflowers
- apple cider vinegar or good-quality white wine vinegar

Place the flowers in a colander with a dishcloth over the top and leave for half an hour so any pollen bugs will crawl off. Fill a large bottle with the flowers and cover with vinegar. Poke with a chopstick to remove any air bubbles. Seal and leave in a cool dark place for 6 weeks, shaking occasionally. Strain into a sterilised bottle.

SPRING BOOSTER SALAD

Spring is a time when many people suffer from hay fever. Try this salad to reduce the symptoms by boosting the immune system. It is also a good salad to cleanse and tone our bodies after the heavier foods of winter.

Dressing

- ½ cup mixed parsley, spearmint, coriander, lovage and fennel leaves
- 4 tablespoons olive oil
- 1 tablespoon balsamic vinegar
- ½ teaspoon freshly grated horseradish
- 1 teaspoon honey
- salt to taste

Salad

- a large bowl of a selection of young dandelion, rocket, nasturtium, clover, alfalfa, gotu kola, sorrel and watercress leaves
- handful of alfalfa, nasturtium and calendula petals
- 3 tablespoons pumpkin seeds
- 3 tablespoons sunflower seeds

For the dressing, finely chop the herbs in a blender. Add the remaining ingredients and blend until smooth.

For the salad, wash and dry the leaves and petals well. Mix together in a bowl. Toast the seeds in a dry frying pan for 2–3 minutes until just starting to brown. Sprinkle over the salad and drizzle with the dressing. Toss and serve.

GREEN HERB SOUP

Serve this soup cold on a hot summer's day and taste the herbs' fresh flavours punching through. For extra pizzazz, add ice cubes with frozen chive flowers. You can substitute the leafy greens I've used here with what is available.

- 2 tablespoons olive oil
- 1 medium onion, peeled and sliced
- 4 garlic cloves, peeled and diced
- 2 teaspoons fresh oregano, chopped
- 1 tablespoon dried mint
- 1 litre vegetable or chicken stock
- 250 g baby spinach leaves
- 250 g watercress leaves
- 100 g sorrel leaves
- 1 cup fresh parsley
- 1 cup fresh coriander
- salt and black pepper to taste
- squeeze of lime juice
- pecorino cheese shavings

Herbed yoghurt
- 1 tablespoon fresh coriander
- 1 tablespoon fresh mint
- 1 tablespoon fresh parsley
- 1 tablespoon fresh chives
- 1 cup of yoghurt

Heat the olive oil in a heavy-bottomed saucepan over low heat. Sauté the onion for about 5 minutes. Add the garlic, oregano and mint and sauté for a further 5 minutes, stirring occasionally. Add the stock and bring to a boil. Add the spinach, watercress, sorrel, parsley and coriander. Reduce to a simmer and cook for 5 minutes. Remove from the heat and use a stick blender to blitz until smooth. Add salt and pepper to taste. Add the lime juice. Leave to cool and then place in the fridge to chill. Taste again for seasoning once chilled as it might need further salt and pepper.

To make the herbed yoghurt, blend the fresh herbs until very finely chopped and mix with the yoghurt. Serve the soup cold with the yoghurt swirled on the surface and more on the side for diners to add as they please. Top with pecorino cheese shavings.

MEDITERRANEAN SALAD

This delicious salad is a meal on its own. Serve it in the garden with a crusty loaf of bread and elderflower cordial. Coriander goes to seed so quickly and abundantly. Instead of leaving them to dry completely, pick the seeds when they are fresh. The zesty green balls can be crushed to a paste that tastes of both the dried seed and fresh leaf.

To make the dressing, dry roast the coriander and cumin seeds in a hot pan until fragrant and then grind finely. (If you are using fresh coriander seeds, skip the roasting.) Place the ground seeds in a blender with all the remaining ingredients and blitz until smooth.

For the salad, simmer the lentils until cooked but still firm. Drain and leave to cool. Steam the broccoli until just cooked and leave to cool. Mix the lentils and broccoli together with the dressing. Mix the carrot, pawpaw, pumpkin seeds and raisins together. Toss the mixture gently into the broccoli and lentils. Top with edible flowers.

Dressing
- ½ teaspoon coriander seeds
- 1 teaspoon cumin seeds
- ½ cup mixed coriander and parsley
- 4 tablespoons olive oil
- 1 teaspoon finely grated ginger
- 1 small hot chilli, deseeded and sliced
- 1 clove garlic, finely chopped
- 1 tablespoon paprika
- 1 tablespoon lemon juice
- salt to taste

Salad
- ½ cup brown lentils
- 2 cups broccoli florets
- ½ cup grated carrot
- ½ cup grated green pawpaw
- 1 tablespoon pumpkin seeds
- 1 tablespoon raisins
- edible flowers

ELDERFLOWER SYRUP

This syrup is delicious added to desserts. It can also be mixed with soda water for a fresh summery drink: pour about 2 tablespoons into a glass, top with soda water, add a squeeze of lemon or lime juice and crushed ice.

- about 30 heads of elderflowers
- 2 litres water
- 3 lemons, thinly sliced
- 1 kg sugar

Place the flowers in a colander with a dishcloth over the top so any bugs will crawl off. Remove all the thicker green stems (don't worry about the smaller flower-bearing stems). In a large pot, bring the water, lemons and sugar to a boil, stirring until the sugar dissolves.

Add the flowers, cover the pot with a lid and remove from the heat. Leave in a cool place for 2 days. Strain and then heat the liquid. As soon as it is simmering, remove from the heat, pour into sterilised bottles and seal. This syrup will keep for up to a year.

LEMON AND MINT ZINGER

Chopping up the lemon skins adds a zesty flavour to this wonderfully fresh and healthy drink. It is especially good served on a hot summer's day or accompanying a spicy meal. Add gin or rum to transform it into a refreshing cocktail.

- 3 large lemons
- 1 cup mint leaves
- ice
- 1 litre water
- runny honey to taste

Wash the lemons well. Squeeze the juice into a bowl and roughly chop the skins. Add the skins, mint and ice to a blender and pulse until well chopped. Cover with water, leave to sit for 10 minutes, then strain, discarding the skins and ice. Mix the lemon-skin water with the lemon juice. Add honey to taste. Pour into individual glasses and top with crushed ice and a mint leaf.

WINTER WARMER SOUP

I can't make it through winter without butternut – it is such perfect comfort food for a chilly night. Combined with carrots and sweet potatoes in this spicy and warming soup, it will increase your circulation. This soup contains plenty of garlic to keep your immune system healthy.

Preheat the oven to 200 °C. Toss the onion, butternut, carrots and sweet potatoes in a bowl with the olive oil, ginger, seeds and chilli. Place the rosemary sprigs on a baking tray and spread the vegetables on top, scraping out all the oil and spices from the bowl. Slice the top off the head of garlic, exposing the cloves. Place the head on a piece of aluminium foil. Drizzle with olive oil and then enclose it in the foil. Place it in the centre of the vegetables. Roast the vegetables for 40–50 minutes, turning them once (except the garlic), until cooked.

Remove from the oven and leave to cool for about 10 minutes. Remove the garlic from the aluminium foil and squeeze the individual cloves into a heavy-bottomed pot. Add the vegetables, discarding the woody rosemary stems, scraping the rest into the pot. Add the stock and, using a stick blender, whizz until smooth. Add salt and pepper to taste and simmer for 10 minutes. Serve topped with the fresh parsley and coriander and a drizzle of yoghurt.

- 4 red onions, roughly sliced
- 2 kg evenly sliced butternut, carrots and sweet potatoes
- olive oil
- 2 cm ginger, sliced
- ½ tablespoon coriander seeds
- 1 teaspoon cumin seeds
- 1 dried hot chilli, crushed
- 3 long sprigs fresh rosemary
- 1 head of garlic
- 1 litre chicken or vegetable stock
- salt and pepper to taste
- ½ cup fresh parsley, chopped
- ½ cup fresh coriander, chopped
- yoghurt

HANGOVER KICKER

- 1 tablespoon sea salt
- ½ cup mixed alfalfa, borage and thyme leaves and flowers
- ¼ cup dried dandelion, milk thistle and gotu kola leaves
- 1 teaspoon ground fennel seeds
- 1 teaspoon ground coriander seeds
- ½ teaspoon dried ginger
- 2 tablespoons bicarbonate of soda
- 5 tablespoons powdered glucose (available from pharmacies)

This cleansing drink is perfect to rid the body of toxins and make you feel better after partying a wee bit hard. It is also good to drink if you have been eating too much rich food and need to cleanse your body.

Heat the oven to 80°C. Mix the sea salt and fresh herbs in a blender and whizz until finely chopped. Spread out on a baking tray and pop in the oven. Turn the heat off and leave for 2 hours to dry. Mix with the remaining ingredients and store in an airtight container.

To drink, place a tablespoon or two in an infuser spoon and add warm water. Infuse for 5 minutes before drinking.

ROSEMARY AND LIME SALT

Rosemary bushes can grow huge. This recipe takes advantage of the herb's abundance. Use as seasoning on just about anything where you would normally use salt.

- 1 large bunch rosemary
- 2 cups sea salt
- 1 tablespoon lime zest

Strip the rosemary off its stems and blend until finely chopped. Add the salt and blend further until it is smooth and green. Mix in the lime zest, pour the mixture into a bottle and seal.

HERB AND HONEY PURÉE

This is a quick and easy way to make a herbal drink. It lasts for about two weeks in the fridge.

- 2 tablespoons herbs such as lemon balm, mint or anise hyssop
- ½ cup runny honey
- juice of ½ lemon

Blend the herbs in a processor and add the honey and lemon juice. Blitz until smooth.

To drink, dilute with water to taste.

HERB BUTTER

Using herb butter is a great way to quickly add flavour to a meal. You can mix a few herbs together or use only one herb. Instead of making a herb-butter log, use chocolate moulds (available from specialist baking shops) to make individual herb butters for a dinner party.

Chop the herbs finely and cream together with the lemon zest and butter until evenly mixed. Mix in salt to taste. Turn onto some wax paper and use the paper to roll it into a log shape. Seal the ends of the wax paper and freeze. To use, cut off a slice and place on top of hot food or use for cooking.

- ½ cup of a selection of herbs such as chives, coriander, parsley or lemon thyme
- 1 teaspoon lemon zest
- 500 g unsalted butter at room temperature
- sea salt to taste

HERB OIL

Different from infused oil, this herbal oil is blend of herbs and oil, resulting in a bright green, herb-flavoured oil that can be drizzled onto soups or vegetables. I have used chives for this recipe but other herbs that also work are parsley, coriander and basil.

Blanch the chives for about 10 seconds in salted, boiling water and then quickly dunk them in iced water to cool off. Twist them in a tea towel to dry and chop roughly. Pop the chopped chives into a blender with the oil and blitz until smooth. Pour into an oil or coffee filter fitted over a jug or bowl and leave to drip for 2–3 hours or overnight. Add salt to taste. Best used fresh, but does keep for about a week in the fridge.

- ½ cup chives
- ¾ cup olive oil
- salt to taste

KIWI FRUIT ICED LOLLIES

Super refreshing on a hot summer's day, these cool kiwi fruit iced lollies are flavoured with a mixture of lemon herbs, plus the added touch of strawberries. Experiment with other flavour combinations – mango, mint and melon is also delicious.

Stir the sugar and water together in a pot over low heat until the sugar has dissolved. Bring to the boil and add the salt and herbs. Simmer for 5 minutes. Remove from the heat and leave to cool. Strain and mix the syrup with the kiwi fruit purée. Divide the strawberries amongst the iced lolly moulds and pour the kiwi mixture on top, leaving about 5 mm on top for expansion. Use a lolly stick to stir the strawberries evenly through the kiwi fruit. Insert the sticks and freeze until solid. If there is any mixture left over, freeze it in ice trays.

- ½ cup white sugar
- ½ cup water
- pinch of salt
- 2 sprigs lemon thyme
- 2 sprigs lemon verbena
- 1 stalk lemon grass, crushed
- 2 cups puréed kiwi fruit
- 1 cup strawberries, cut into slices

ICED HERB BOWL

This simple technique creates a bowl made from ice, perfect for serving any cold dish – from ice cream to oysters. You will need two bowls that fit one inside another. I like to use glass as you can see where you are placing the flowers and herbs, but any kind of bowl will do. Boiling the water first removes any air bubbles and makes the ice bowl clearer.

Pick a selection of herbs and edible flowers that will suit what you are serving. Spread some crushed ice on the bottom of the larger bowl and place some herbs and flowers on top. Cover with a little more crushed ice and place the smaller bowl on top, fitting it inside the larger one and centering it. Place something inside the smaller bowl to hold it in place (a bag of frozen vegetables will do). Use pieces of tape across the top of the bowls to hold them together in position. Pour enough cooled boiled water into the gap between the bowls until it reaches just above the bottom of the inside bowl, and freeze. Remove from the freezer and add another layer of flowers and herbs, sprinkling crushed ice on them to keep them in place. Add more water and freeze again. Repeat doing this until it is frozen all the way to the top.

Take it out of the freezer and remove the tape and frozen vegetables. Pour a little warm water into the inner bowl, swirling it until the bowl is loose enough to remove. Remove the outside bowl using warm water again. Place it on cling wrap and cover loosely. Refreeze until you are ready to use it. Serve it on a tray with a lip so it won't make a mess when it melts.

- selection of herbs and flowers
- 2 nested bowls
- crushed ice
- boiled water, cooled

THEREIN LIES THE RUB

Rubs are a wonderful way to make use of abundant herbs in midsummer. The dried olives and tomatoes add a flavour punch to this mix and it works particularly well on roasted vegetables. Don't use sundried tomatoes stored in oil, otherwise it will make the mixture too oily. If that is all you can find, drain them from the oil, rinse and pat dry.

- 1 tablespoon each dried thyme, dried oregano, crushed dried rosemary
- 1 teaspoon each dried marjoram, dried basil
- 1 tablespoon dried olives, finely chopped
- 1 tablespoon sundried tomatoes, finely chopped
- 2 teaspoons sea salt
- ½ teaspoon ground pepper
- 1 teaspoon paprika

Mix all ingredients together and seal in an airtight container.

To use, rub on meat, fish or poultry before cooking or mix with olive oil and toss with vegetables before roasting.

PLUM, CHOCOLATE AND MINT SAUCE

I cooked with the King of Chocolate, Willie Harcourt-Cooze, when he was visiting South Africa. He gave me a selection of his 100% cacao cooking chocolate to play with, and this was one of the delicious things I made.

Place the plums and water in a pot, cover and bring to the boil. Turn off heat and leave to cool. Strain plums through a colander into a bowl, pushing with a spoon to get as much of the pulp as possible through, leaving the skins and pips behind. Measure, and for each cup of plum pulp add one cup of sugar. Place in a pot with the mint leaves and bring to the boil. Cook, stirring often, until it has thickened. (Test by dropping a little onto a chilled plate. If it forms a skin when you push it, it's ready.) Remove from the heat and quickly stir in the chocolate and brandy. Decant into sterilised jars.

- 1 kg plums
- 1 cup water
- sugar
- ½ cup fresh mint leaves
- ½ cup grated good-quality dark chocolate
- ½ cup brandy

Around The House *Household cleaning, pest repelling, air freshening*

Progress in household cleaning products and aids has played a large role in liberating women from the drudgery of cleaning and washing. However, many of these chemical cleaners and products damage the environment and our health. I am not advocating we return completely to Granny's way of life, but we can go back to using natural ingredients to create effective cleaning aids that won't harm the environment.

ALL-PURPOSE CLEANER

This fresh-smelling cleaner can be used to clean bathroom and kitchen surfaces without any harsh chemicals.

- ¼ teaspoon thyme-, tea tree- and rosemary-infused oil
- a few drops each of citronella, bergamot, rosemary and tea tree essential oils
- 2 teaspoons vinegar
- 2 cups water
- 1 tablespoon bicarbonate of soda

Mix the oils in a spray bottle. Add the vinegar, water and bicarbonate of soda. Shake well. Spray over surfaces and leave for 10 minutes before wiping clean.

MOULD CLEANSER

Mould in the home can cause allergies, asthma and other health problems. Moulds usually grow in high-moisture areas such as the shower or in corners above the stove. Use this mixture to disinfect and remove mould. If you have grout that has become discoloured, this will also whiten it.

Mix the ingredients together in a spray bottle and shake well. Spray the mouldy surface and leave for 24 hours before rinsing off.

- 2 tablespoons borax
- 1 cup white vinegar
- 2 cups hot water
- ¼ teaspoon each tea tree and rosemary essential oils

AIR FRESHENER

This fresh-smelling mixture will clear a room of tobacco smoke, cooking odours and other stale smells.

Mix the ingredients together well and seal in a sterilised bottle. Heat a few teaspoons of the mixture in an oil burner as required.

- 1½ cups apple cider vinegar
- 1 tablespoon rose water
- 3 drops each coriander, eucalyptus, tea tree and lavender essential oils

MOTH-REPELLENT MIX

I really don't like the smell of mothballs, either on my clothes or on someone else's. But people use them for a reason: clothes-eating moths certainly hate them. African wormwood, feverfew, lemon verbena, rosemary, rose-scented pelargonium and lavender leave your clothes smelling fresh and sweet. In addition, it takes care of another problem we all have in our homes – the sock thief. Somehow, no matter how careful we are, we land up with single socks. Now you have a use for them. Stuff them with herbs and tie a knot in the top to create a no-fuss herb sachet.

Hang the herbs to dry for a few days. Remove the leaves from the stems and mix them together. For every cup of herbs, add ½ teaspoon cloves and 1 teaspoon dried citrus peel. Divide the mixture amongst single socks or old stockings and tie the tops closed. Toss in amongst your jerseys.

- equal quantities of strong-smelling herbs
- cloves
- dried lemon, lime or other citrus peel

SCOURING PASTE

This fresh-smelling scouring paste can be used to clean stubborn dirt and stains from baths, counters and toilets. Rub it on and leave it for 10 minutes before starting to scrub. It also works well to clean really grubby hands and feet!

- 1 cup water
- 1 cup natural soap (see page 184)
- small block pumice
- 1 tablespoon dried lavender flowers, crushed
- 10 drops tea tree essential oil
- 10 drops lemon grass essential oil
- 10 drops lavender essential oil

Heat the water until it is boiling and add the soap. Reduce the heat and stir until all the soap has melted. Remove from the heat and stir often as it cools until it thickens into a smooth, thick paste. If necessary add more water. Wrap the pumice in an old cloth and crush it using a hammer. Use a pestle and mortar to grind it to a fine powder. Add the pumice powder, the lavender flowers and the essential oils to the soap mix and stir through. Store in a sealed, wide-mouthed container.

FRIDGE FRESHENER

This mixture keeps a fridge smelling fresh. The bicarbonate of soda and salt absorb odours and the herbs give off their scent. Replace it once a month.

- 1 cup bicarbonate of soda
- 2 tablespoons salt
- 1 cup chopped mixed herbs such as mint, pelargonium, lavender and rosemary

Mix all the ingredients in a bowl and place it on a shelf in the fridge. Stir it once a week.

SOAPWORT WASHING LIQUID

Soapwort creates a very gentle washing liquid, ideal for delicate clothing.

- 1 litre water
- 2 cups chopped soapwort leaves and stems
- 1 cup lavender flowers
- 10 drops lavender essential oil

Bring the water to the boil, add the soapwort and lavender and cover. Reduce the heat and simmer for about 15 minutes. Remove from heat and cool. Strain and add the essential oil. Use immediately or store in the fridge for up to a week.

INSECT REPELLENT

The herbs in this candle repel insects such as flies and mosquitoes. If you don't want the hassle of making a candle, simply blend the chopped herbs, dry them as described below and mix with some crushed frankincense (available at Indian spice shops). Store the mixture in a sealed container in the freezer to prevent it going mouldy. To use, place a large pinch in the top of an oil burner and light the candle. There is no need to add any oil as the natural oils in the leaves will be sufficient.

Preheat the oven to 100 °C. Chop the herbs finely in a blender and spread out on a baking tray. Dry in the oven for 4 hours and then turn off the heat and leave overnight. Place the herbs in the top of a double boiler or bowl set over simmering water. Add sufficient beeswax to fill your mould. Heat until the beeswax has melted completely. Remove from the heat, add the essential oils and stir until well mixed. Slowly pour into the mould, tapping the sides to remove any air bubbles, and leave to set.

- 1 bowl mixed fresh lemon balm, lemon grass, lemon verbena, tansy, African wormwood, feverfew, citronella pelargonium and rose-scented pelargonium
- beeswax, grated or chopped
- 10 drops each citronella oil, lemon grass and rose-scented pelargonium essential oils
- 1 candle wick and mould (available from craft shops)

Delicious Beauty Care *Skin, body and hair care recipes*

Using homemade herbal beauty products is empowering and also contributes to our general health and wellbeing. Many commercial products contain harmful preservatives and chemicals. Making our own products is like making food from scratch – we know exactly what has gone into it. The recipes in this section should be used as a jumping off point for you to experiment with your own variations.

WRINKLE AND CRINKLE BALM

Use this gentle balm around your eyes and mouth and on your neck to reduce the appearance of fine lines.

- 4 tablespoons olive oil
- 2 tablespoons each gotu kola leaves and evening primrose flowers and leaves
- 4 tablespoons calendula flowers and leaves
- 10 g beeswax
- 2 tablespoons jojoba oil
- 1 tablespoon rose hip oil
- contents of 2 capsules vitamin E oil
- 10 drops each lavender, chamomile and rose essential oils

Mix the olive oil and herbs together in a double boiler or a bowl set over water and heat gently for about 3 hours. Strain the mixture into a bowl or jug. Wipe the double boiler with paper towel and heat the beeswax until it is just melted. Add all the oils, except the essential oils, stirring until mixed. Remove from the heat and add the essential oils. Decant into sterilised jars.

PERFUME OIL OR BALM

I am very sensitive to most perfumes and often find them overwhelming – especially if someone sprays a perfume on in excess. A number of years ago I discovered an Indian perfume balm that was subtly scented with natural essential oils. I experimented with making my own and since then I only use essential oils as perfume. This can be made into a balm, to dab on wrists and neck, or an oil, which can be added to your hair-rinse water, resulting in a very subtle scent. This recipe uses my favorite mix of essential oils – adapt it using yours.

Perfume balm

Mix the beeswax and the infused oil in a double boiler or bowl set over simmering water. Heat until the beeswax melts, stirring occasionally. Add the essential oils, remove from the heat and decant into sterilised jars.

Perfume oil

Mix the infused oil with the essential oils and decant into a sterilised bottle. Add a few drops to the rinsing water after washing and conditioning your hair.

Perfume balm
- 15 g beeswax, grated or chopped
- ½ cup rose-scented pelargonium infused oil
- 10 drops each vanilla, neroli and grapefruit essential oils

Perfume oil
- ½ cup rose-scented pelargonium infused oil
- 10 drops each vanilla, neroli and grapefruit essential oils

HAIR-STRENGTHENING JUICE

Drink a cup of this juice three times a day to increase hair growth and to strengthen hair, preventing split ends.

- 1 cup water
- ½ cup mixed fresh alfalfa, gotu kola, nettle, parsley, rosemary and watercress leaves
- 3 heads of dark green or red lettuce (the darker the lettuce, the more nutrients it contains)
- 10 large carrots

Boil the water and pour over the herbs. Leave to infuse for 10–15 minutes. Strain and cool. Juice the lettuces and carrots. It is easier to juice the lettuce if you wrap the leaves around the pieces of carrot. Mix the juice with the herbal infusion.

PIMPLE FIXER

- 2 ½ cups apple cider vinegar
- ¼ cup calendula petals
- ¼ cup rosemary leaves
- ¼ cup tea tree leaves
- ¼ cup parsley

Tea tree oil on its own will sort out most pimples: simply dab some undiluted tea tree oil onto the spot overnight. This goes against the usual rule of not using undiluted essential oils on your skin – but in the case of a painful pimple this really works. The following recipe is for long-term prevention of pimples.

Mix the vinegar and herbs together in a bottle. Seal and leave to infuse in a warm sunny spot for 6 weeks, shaking regularly. Strain and pour into a sterilised bottle. Add a few dollops to the rinsing water when washing your face.

TONER FOR OILY SKIN

This toner helps treat pimples and acne by cleaning excess oil and tightening the pores. To use, gently wipe over your skin and follow with moisturiser.

Mix the ingredients in a dark glass bottle and shake well to combine. Wait 24 hours before using.

- 1 cup witch hazel
- 3 drops lemon grass essential oil
- 3 drops tea tree essential oil
- 1 drop rose-scented pelargonium essential oil

FRANKINCENSE AND ROSE SKIN CREAM

Frankincense adds its toning and restorative qualities to this luxurious cream. It can be used as a body and face cream.

Heat the beeswax and cocoa butter in a double boiler or a bowl set over water until melted. Add the calendula oil and stir until well mixed. Remove from heat and whisk often until it has cooled down and thickened. (To speed up cooling, place the pot in a bowl of iced water but make sure no water splashes into the mixture.) Add the basic cream, whisking constantly until smooth. Mix in the essential oils. Store in a wide-mouthed jar. This cream will keep for three or four months.

- 10 g beeswax, grated or chopped
- 30 g cocoa butter
- 1 tablespoon calendula oil
- ½ cup basic cream (page 182)
- 10 drops rose essential oil
- 10 drops frankincense essential oil

ANTI-BUG BALM

Lemon balm, lemon grass, lemon verbena, citronella pelargonium and rose-scented pelargonium not only smell good – they repel mosquitoes. Rub this on your skin to keep pesky mozzies away.

Mix the olive oil and herbs together in a double boiler or a bowl set over water and heat gently for about 3 hours. Strain the mixture into a bowl or jug. Wipe the double boiler with paper towel and heat the beeswax until it is just melted. Add the infused oil and stir until it is mixed. Remove from the heat, stir in the essential oils and transfer to a storage jar.

- 1 cup olive oil
- 2 cups chopped, anti-mozzie herbs
- 25 g beeswax, grated
- 5 drops each lavender and lemon grass essential oils

EDIBLE BODY BALM

- 3 tablespoons olive oil
- ¼ cup apple mint, chopped
- 30 g cocoa butter
- 10 g beeswax, grated or chopped
- 1 teaspoon pure dark chocolate, grated
- ½ teaspoon stevia powder
- 10 drops peppermint essential oil

This delectable chocolate and mint balm is a romantic anniversary or Valentine's Day gift.

Mix the olive oil and mint together in a double boiler or a bowl set over water and heat gently for about 3 hours. Strain the mixture into a bowl or jug. Wipe the double boiler with paper towel and melt the cocoa butter and beeswax. Add the infused mint oil and chocolate and stir until mixed. Remove from the heat. Stir in the stevia powder and peppermint essential oil and mix well. Decant into sterilised containers.

YUMMY MUMMY BODY BUTTER

With gotu kola to restore collagen and elasticity, and evening primrose, calendula and vitamin E oil to reduce scarring, this body butter will help reduce stretch marks and keep skin toned.

- 10 g beeswax, grated or chopped
- 2 tablespoons cocoa butter
- 2 teaspoons wheat germ oil
- contents of 4 capsules vitamin E oil
- contents of 4 capsules evening primrose oil
- 2 teaspoons calendula oil
- 2 tablespoons gotu kola infusion
- 1 cup basic cream (page 182)
- ⅛ teaspoon borax powder
- 5 drops each lavender, mandarin and neroli essential oils

Heat the beeswax in a double boiler or a bowl set over water until it is just melted. Add the cocoa butter and melt. Add the wheat germ, vitamin E, evening primrose and calendula oils and stir until well mixed. In a separate bowl, whisk the gotu kola infusion, basic cream and borax powder together and heat until it reaches about 75 °C. Add the beeswax mixture in a slow drizzle, stirring constantly, until smooth. Remove from heat. Add the essential oils. As it cools, stir it often, whipping it into a creamy texture. Store in a wide-mouthed jar. This body butter will keep for three or four months.

EXFOLIATING BODY SCRUB

Polish, cleanse and soften your skin with this delicious body scrub. It is especially good for dry, rough skin. To use, rub it into wet skin in a circular motion, gently exfoliating. Rinse it off afterwards. The calendula oil moisturises and encourages healthy skin regeneration.

- 10 g beeswax, grated or chopped
- 30 g cocoa butter
- 1 tablespoon shea butter
- 2 tablespoons calendula-infused oil
- ½ cup basic cream (page 182)
- 5 drops each sandalwood, cedar wood, frankincense and rose essential oils
- 2 tablespoons pumice stone

Heat the beeswax in a double boiler or a bowl set over water until it is just melted. Add the cocoa butter and shea butter and melt. Add the calendula oil and stir until well mixed. Remove the mixture from heat and leave to cool, stirring often and whipping it as it cools. Add the basic cream, whipping constantly until smooth. Add the essential oils. Wrap the pumice in an old cloth and bash it with a hammer to break it into small pieces. Crush it to a fine powder in a pestle and mortar and stir into the mixture. Store in a wide-mouthed jar. This scrub will keep for three or four months.

LAVENDER AND NEROLI BODY POWDER

This body powder is ideal for keeping cool in hot weather. It keeps skin fresh and protects against chafing. To prevent and treat fungal problems, add tea tree essential oil.

- ½ cup cornflour
- 10 drops each lavender, neroli and spearmint essential oils

Spread the cornflour evenly onto a flat tray or plate. Sprinkle the essential oils over the cornflour, whisking to distribute them evenly. Leave to dry before decanting into a clean shaker bottle.

LOOFAH SCRUBBER SOAP

Scrubbing with loofah exfoliates the skin, which is especially good for elbows and grubby gardening feet. As you use this soap, so more loofah is exposed. The soap ingredients can be found at specialist soap-making shops, or online.

- loofah sponge
- soap moulds
- 200 g pure glycerine soap
- 2 tablespoons aloe vera gel
- 20 drops peppermint essential oil
- soap colouring of your choice

Soak the loofah in water until it has expanded. Cut it into discs or pieces that fit tightly into the moulds and place one in each mould. Slice the glycerine soap into small chunks and put them in a double boiler or a bowl set over water and heat until just melted. Don't overheat the soap and let it boil or it will go cloudy. Remove from the heat and stir in the gel, essential oil and colouring. (Start with a few drops of colouring and stir it in. Repeat until you are happy with the colour.) Pour the soap into the moulds, giving each one a couple of taps to make sure the soap fills all the gaps in the loofah. Leave to set. If the loofah floats to the top as you pour, only fill the mould halfway. Leave it to set a little, then pour in the remainder of the melted soap. You can also use this method to create layers of colour. To retain the peppermint scent, wrap the soaps in pure cotton cloth.

SHINY SHAMPOO

- 2 cups water
- 3 tablespoons soapwort leaves and stems
- 3 tablespoons nettle root
- 3 tablespoons rosemary

This shampoo contains herbs to stimulate hair growth and help keep your hair healthy and shiny.

Bring the water to the boil, add the herbs and cover. Reduce the heat and simmer for about 15 minutes. Remove from the heat and leave to cool. Strain the mixture and store it in the fridge. It will keep for a week.

ROSEMARY AND PEPPERMINT SHAMPOO SOAP

This is a great soap to take travelling – a small amount goes a long way and there is no need to carry a whole bottle of shampoo with you.

Slice the soap into chunks and heat them in a double boiler or a bowl set over water until just melted. Remove from the heat and stir in the oils and herbs. Pour the soap into the moulds and leave to set. To retain the essential oils, wrap the soap in pure cotton cloth. If you want the herbs to be evenly distributed throughout the soap, pour a first layer and leave it to set. Remelt the soap mixture and pour a second layer on top. Repeat until the mould is full.

- 200 g pure soap
- 1 teaspoon lavender- and rosemary-infused olive oil
- ½ teaspoon rosemary essential oil
- ½ teaspoon peppermint essential oil
- 2 tablespoons dried rosemary and lemon verbena, crushed
- soap moulds

CONDITIONER FOR DRY HAIR

Mix the olive oil and herbs together in a double boiler or a bowl set over water and heat gently for about 3 hours. Strain the mixture into a bowl or jug. Wipe the double boiler and heat the shea butter until it is just melted. Stir in all the oils apart from the essential oil. Remove from the heat and cool, stirring often, until the mixture thickens. Add the basic cream, whipping constantly, until smooth. Add the essential oil and decant into a sterilised container.

Comb this rich conditioner through wet hair after washing and leave it on for a couple of hours before rinsing out.

- 1 cup olive oil
- 1 cup mixed alfalfa, gotu kola, parsley, nettle and rosemary leaves
- 2 tablespoons shea butter
- 4 tablespoons coconut oil
- 4 tablespoons jojoba oil
- 1 cup basic cream (page 182)
- 10 drops rosemary essential oil

ANTI-FUNKY FOOT BALM

- 30 g cocoa butter
- 30 g beeswax, grated or chopped
- 2 tablespoons shea butter
- 2 tablespoons almond oil
- 20 drops peppermint essential oil
- 5 drops tea tree oil
- 10 drops lavender essential oil
- 1 cup basic cream (page 182)

This is a refreshing moisturising balm for feet, with tea tree oil to keep any fungal baddies away.

Melt the cocoa butter, beeswax and shea butter in a double boiler or a bowl set over simmering water. Stir in the oils and mix well. Remove from the heat and whip the mixture often as it cools. Whip in the basic cream until smooth and decant into sterilised containers.

TINTED LIP BALM FOR DRY LIPS

This moisturising lip balm keeps lips moist while adding a blush of colour. Peppermint oil gives it a zingle. I use natural powder to colour my lip gloss, available from Inthusiasm.

- 15 g beeswax
- 10 g cocoa butter
- 1 tablespoon wheat germ oil
- 1 tablespoon jojoba oil
- 1 tablespoon olive oil
- contents of 1 capsule vitamin E oil
- 10 drops of peppermint essential oil
- your favourite lipstick or natural powder to colour

Melt the beeswax and cocoa butter in a double boiler or bowl set over simmering water. Add all the oils and the lipstick or powder, stirring until it is well mixed and has reached the colour intensity you like. Remove from the heat and decant into small sterilised jars.

SAGE AND PEPPERMINT TOOTHPASTE

- ¼ cup watercress
- ¼ cup sage
- ¼ cup nettle
- ½ cup coconut oil
- bicarbonate of soda (enough to thicken)
- ¼ teaspoon stevia powder
- 20 drops myrrh essential oil (available from health shops)
- 8 drops clove essential oil (optional)
- 25–30 drops peppermint essential oil

Many commercial toothpastes contain a scary list of ingredients, such as foaming agents, fluoride, flavouring and colouring. Even if toothpaste is not swallowed, these ingredients are absorbed through the mucous membranes of our mouth. All the more reason to make our own. This toothpaste contains watercress and nettle to strengthen teeth, plus baking soda, which has been used for centuries to clean teeth. Stevia sweetens the toothpaste and it helps reduce bacteria. Add the clove oil if you have sensitive teeth, as it is a mild anaesthetic and will reduce tooth sensitivity. The tincture of myrrh keeps gums healthy, and the peppermint not only tastes good, it freshens the mouth.

Mix the herbs and coconut oil together in a double boiler or a bowl set over water and heat gently for about 3 hours. Strain the mixture into a bowl or jug. Add the bicarbonate of soda a tablespoon at a time and stir with a fork until it forms a thick paste. Stir in the stevia powder, and myrrh and clove (if using) essential oils. Add the peppermint essential oil to taste. Decant into a storage jar. Keep enough for a few days in your bathroom and store the remainder in your fridge.

LEMON FRESH ROSE PETAL BATH BOMBS

These are great gifts to make. The basic recipe is the same – just add whichever herbs and essential oils you want to use. If you want to add colouring, do it at the same time as mixing in the oil. (Don't use too much otherwise you will stain your bath!)

- 2 tablespoons citric acid
- 3 tablespoons bicarbonate of soda
- 2 tablespoons mixed dried lemon verbena, lemon balm and lemon grass, chopped finely
- 1 tablespoon dried rose petals
- 2 teaspoons jojoba oil
- 10 drops lemon grass essential oil

Mix the citric acid and bicarbonate of soda together using a whisk to make sure they are well incorporated. Add the dried lemon herbs and rose petals and mix through. Mix the jojoba oil and lemon grass essential oil together and sprinkle evenly over the surface, whisking as you sprinkle. Mix together well. Press firmly into biscuit cutters, soap moulds or ice cube trays to shape the mixture, and leave to dry.

ROSE-SCENTED PELARGONIUM HAND CREAM

This fragrant and smoothing cream nourishes dry, hard-working hands.

- 10 g beeswax
- 2 tablespoons shea butter
- 1 tablespoon almond oil
- 1 cup basic cream (page 182)
- 3 tablespoons rose water
- 1 vitamin C tablet (500 mg), crushed
- 15 drops rose-scented pelargonium essential oil

Melt the beeswax and shea butter in a double boiler or bowl set over simmering water. Add the almond oil, stirring until well mixed. Remove from the heat and leave to cool, stirring often, until it has thickened. In a separate bowl, mix the basic cream, rose water and vitamin C together. Add the cream mixture to the beeswax and oil, whipping until it is smooth. Stir in the essential oil and decant into a sterilised jar.

EVENING PRIMROSE FACE CREAM

This rich face cream is a good night moisturiser for dry skins.

Melt the beeswax in a double boiler or bowl set over simmering water. Add the jojoba, rosehip, vitamin E and evening primrose oils and stir together until well mixed. Remove from the heat and leave to cool, stirring until the mixture thickens. Mix the basic cream, rose water and vitamin C together in a separate bowl. Add the cream mixture to the beeswax and oil, whipping until it is smooth. Stir in the essential oils and decant into a sterilised jar.

- 10 g beeswax
- 1 tablespoon jojoba oil
- 1 tablespoon rosehip oil
- contents of 3 capsules vitamin E oil
- contents of 3 capsules evening primrose oil
- ½ cup basic cream (page 182)
- 3 tablespoons rose water
- 1 vitamin C tablet (500 mg), crushed
- 8 drops lavender essential oil
- 4 drops chamomile essential oil

PUFFY EYE TREATMENT

Alternating a hot and cold compress on any inflammation will help reduce it and make it feel better. This method works well to treat tired and puffy eyes.

Cover the herbs with the just-boiled water and leave to infuse for 10–15 minutes. Strain. Divide the mixture between two bowls. Place one in the freezer for half an hour. Just before treating your eyes, heat the contents of the second bowl in the microwave or on the stove until it is hot. (Don't make it so hot that you will burn your eyes, but hotter than lukewarm.) Doing one eye at a time and using one sponge per bowl, alternate placing a hot, then cold sponge over your closed eyes, holding it in place for about 20 seconds at a time. Continue until the hot mixture cools down. Pat your eyes dry and dab cooling eye gel around your eyes (see the recipe on page 228).

- ¼ cup cornflowers
- ¼ cup raspberry leaves
- just-boiled water

COOLING EYE GEL

Simple slices of cucumber placed over the eyes is a classic remedy for cooling and soothing hot puffy eyes. This gel, with anti-inflammatory herbs, takes it one step further. Keep it in the fridge as the cold helps constrict blood vessels and reduce puffiness. It will keep in the fridge for about two weeks.

Place the herbs in a pot and cover with the just-boiled water. Leave to steep for 10–15 minutes and strain. Chop the cucumber and blend it until smooth. Strain it through a sieve, pushing with a spoon to squeeze the juice out. Mix in the aloe vera gel and set aside. Mix the witch hazel and the herb-infused water in a pan and heat until simmering. Add the gelatine and stir until it is dissolved. Remove from the heat and stir in the cucumber and aloe vera mixture. Decant into a sterilised jar and place in the fridge until it has set. Store in the refrigerator for up to two weeks.

- 1 tablespoon each selfheal leaves and flowers, chamomile flowers and raspberry leaves
- 2 cups just-boiled water
- 1 small cucumber
- 2 tablespoons aloe vera gel
- 3 tablespoons witch hazel
- 4 sachets gelatine

GARDENER'S HAND CREAM

At the end of a hard day's gardening, soothe your hardworking hands with this moisturising and healing hand cream.

Melt the beeswax in a double boiler or bowl set over simmering water. Add the oils (except the essential oils), aloe vera gel and honey, and stir together until well mixed. Remove from the heat and stir in the essential oils. Leave to cool, stirring often, until it thickens. Add the basic cream, whisking constantly, until smooth. Decant into a sterilised jar.

- 20 g beeswax
- 1 tablespoon sesame oil
- 2 tablespoons almond oil
- contents of 3 capsules vitamin E oil
- 2 tablespoons calendula-infused oil
- 2 teaspoons aloe vera gel
- 1 teaspoon honey
- 3 drops chamomile essential oil
- 3 drops rose-scented pelargonium essential oil
- 3 drops rosemary essential oil
- 1 cup basic cream (page 182)

GARDENER'S HAND SCRUB

I always used to have the problem of grubby gardening hands. No matter how much I scrubbed there always seemed to be a bit of dirt left, especially around the edges of my nails. That's until a friend recommended using oil. It worked. Simply rubbing olive oil into the corners of my fingers seemed to help. So I experimented a bit more and came up with this recipe.

Mix the olive oil and herbs together in a double boiler or a bowl set over water and heat gently for about 3 hours. Strain the mixture into a bowl or jug. Leave to cool. Wrap the pumice stone in an old cloth and crush with a hammer. Grind to a fine powder, using a pestle and mortar. Mix 2 tablespoons of pumice powder and all the remaining ingredients with the infused oil. Decant into a sterilised jar. To use, wash hands with soap first and then scoop out some scrub and rub it in, massaging it into particularly grubby areas. Preferably leave it on your hands for a few minutes before rinsing off well. Apply the Gardener's hand cream (see the recipe on page 228) after using the hand scrub.

- 2 tablespoons olive oil
- 1 tablespoon soapwort leaves
- 1 tablespoon calendula leaves and petals
- chunk of pumice stone
- ⅔ cup brown sugar
- 1 teaspoon almond oil
- 10 drops rosemary essential oil
- 10 drops lavender essential oil

Common Ailments *Natural herbal remedies*

There are many ways to use herbs effectively to treat common ailments and relieve the discomfort of recurring and long-term conditions. But herbs are also potent and should be treated with respect. For serious conditions, or if you are on existing medication, it is wise to consult a herbalist or general practitioner. The recipes and remedies in this section use herbs in a natural way to improve our health and treat our bodies in a healing and gentle manner.

Coughs, colds, sore throats, flu and hay fever

TEA THYME

- 2 tablespoons chopped fresh thyme
- ½ teaspoon grated ginger
- 1 cup just-boiled water
- 1 teaspoon honey (or more to taste)
- squeeze of lemon juice

Thyme was one of the first herbs I used medicinally and it remains one of my favourite medicinal herbs. If you feel a sore throat or cough coming on, drink a cup of this soothing tea three times a day. It will help stop it in its tracks.

Cover the thyme and ginger with the hot water and leave to infuse for 10–15 minutes. Strain and stir in the honey and lemon juice.

COLD AND FLU TEA

- 1 cup just-boiled water
- 1 teaspoon dried elderflowers
- 1 teaspoon dried peppermint
- 1 teaspoon dried echinacea root
- 1 teaspoon chopped ginger
- 1 tablespoon honey
- squeeze of lemon juice

If you learn to recognise the first warning signs of a cold and immediately start drinking this tea three times a day, it will either make it disappear or reduce it to a milder version. This is a good tea to take before flying to prevent picking up aeroplane bugs. I take a mixed bag of these three herbs with me when I go travelling – you can buy ginger, honey and lemon wherever you are.

Add the hot water to the herbs and ginger and leave to infuse for 10–15 minutes. Strain and mix in the honey and lemon.

CHEST RUBS

When we were kids, Mom always had a strong-smelling chest rub at hand, ready to slather on if we had a cold. When applying a chest rub, don't just rub it onto the front of the chest. Rub it over the rib cage, front and back, particularly the area between the shoulder blades. Chest rubs are best applied at night. Wear an old T-shirt if you don't want it getting onto your pyjamas.

For congested chest and cough

- 15 g beeswax, grated or chopped
- ½ cup infused thyme, yarrow and lemon balm oil
- 5 drops each ginger, eucalyptus, peppermint and fennel essential oils

To soothe asthma and bronchial spasms

- 15 g beeswax, grated or chopped
- ½ cup infused lavender and borage oil
- 5 drops each chamomile, rose-scented pelargonium, ginger and peppermint essential oils

Mix the beeswax and the infused oils in a double boiler or bowl set over simmering water. Heat until the beeswax melts, stirring occasionally. Add the essential oils, remove from the heat and stir often as it cools. Decant into sterilised jars.

ECHINACEA COUGH LOZENGES

Long before I grew echinacea, I bought some cough sweets in Thailand that contained echinacea. They were the best cough sweets I ever used. When I began growing my own echinacea I experimented with different recipes, trying to replicate the flavour. These lozenges not only soothe a sore throat and ease a tickling cough, they also taste good.

- ½ cup dried echinacea root and stems
- 3 ½ cups mixed fresh peppermint, lemon balm, and sage leaves, and thyme and anise hyssop leaves and flowers
- 1 teaspoon lemon zest
- 2 ½ cups just-boiled water
- 1 teaspoon powdered ginger
- 1 cup gum Arabic powder
- 2 tablespoons dried elderberries
- 1 cup hot water
- 2 ½ cups sugar
- ½ cup icing sugar

Mix the herbs and lemon zest together in a pot and add the just-boiled water. Put a lid on the pot and leave to steep until cold. Strain and measure 1 ½ cups of the infusion and stir in the ginger. While the herbs are steeping, crush the gum Arabic and the dried berries together in a pestle and mortar. Add the mixture slowly to the cup of hot water, stirring constantly until the gum Arabic has dissolved and the mixture is syrupy.

In a pot, combine the herb infusion and the gum Arabic mixture with the sugar. Heat, stirring constantly and making sure the sugar dissolves before it comes to the boil. Simmer for 30–40 minutes, stirring constantly as it thickens. It is ready when it boils down to a thick syrup and pulls away from the edges of the pan when you stir. Test for readiness by dropping a small amount into cold water. As soon as it holds together in a hard ball and doesn't spread out in the water, remove from the heat. Pour into a greased baking tray and leave until cool.

Place the icing sugar in a bowl. Using the handle of a teaspoon, scoop out a lozenge-size piece of the mixture and dip it into the icing sugar. Roll it between the palms of your hands to create a ball and place it on a tray dusted with icing sugar. Keep your hands well dusted with icing sugar to prevent sticking, and repeat until finished. Dust the lozenges with icing sugar and twist each one up into an individual paper wrapper. Store in an airtight container.

HONEY AND CINNAMON COUGH SWEETS

These cough sweets have the added benefit of using healing honey instead of sugar. All of these herbs help ease a sore throat and treat a cough. Cinnamon not only adds to the taste, it is rich in antioxidants. If you have a beehive nearby, be warned: the smell of cooking honey attracts bees in droves!

Mix the herbs and water together in a pot and bring to the boil. Remove from the heat, cover and leave to steep until cold. Strain and, in a large pot, combine 1½ cups of honey and ¼ teaspoon of cream of tartar for every 1 cup of herb infusion.

Boil, stirring often, until the mixture reaches 150 °C or until it forms a hard ball when dropped into cold water. (Be careful of burning the honey at this stage – keep stirring and test often.) Remove from the heat and pour into sweet moulds or onto a greased tray. (Score lines on it when it is slightly set.) Leave to cool and set before removing from the moulds or breaking into pieces. Mix the cinnamon and sugar together and sprinkle over the sweets to stop them sticking to each other. Store the sweets in an airtight container in the freezer. (This will keep them hard. If kept at room temperature they will soften and become sticky, as honey absorbs moisture.)

If you want to make sweets that don't need to be kept in the freezer, replace the honey with sugar and cook it to the hard-crack stage.

- ½ cup dried elderflowers
- ½ cup dried echinacea root and stems
- 2 cups mixed fresh peppermint, thyme, lemon balm and sage
- 2 cups water
- honey
- cream of tartar
- ½ teaspoon cinnamon
- 1 tablespoon icing sugar

HAY FEVER G&TEA

This ginger tea keeps hay fever at bay by using herbs with antihistamine and anti-inflammatory properties. If your hay fever is the result of pollen allergies, try to find a locally produced honey as this will help ease the symptoms. Make a batch every morning and take it with you in a flask so you can drink it throughout the day.

- 1 teaspoon grated fresh ginger
- 1 teaspoon fresh nettle leaf
- 1 teaspoon dried chamomile flowers
- 1 teaspoon echinacea root
- 1 teaspoon each fresh yarrow, thyme and oregano
- 2–3 elderflowers or 2 teaspoons dried elderflowers
- handful fresh gooseberries
- just-boiled water
- honey to taste

Place all the herbs and gooseberries in a jug. Pour just-boiled water over and cover. Leave for 15 minutes and strain, squeezing to get all the juice out. Add honey to taste.

ONION, GINGER AND NASTURTIUM SYRUP

Onion syrup is an age-old remedy for both preventing and treating a cold. With the addition of ginger and nasturtium, this syrup packs a powerful punch. Take a teaspoon three to four times a day as soon as you feel the first signs of a cold coming on, and it will either reduce it or chase it away completely.

Mix the onion, ginger and nasturtium in a bottle. Cover with the honey or sugar, put the lid on and leave to sit for a few hours until syrup has formed. Keep in the fridge and make a fresh batch every two days.

- ¼ onion, roughly chopped
- 1 tablespoon ginger, roughly chopped
- ½ cup nasturtium seeds, flowers and leaves, chopped
- 2–3 tablespoons honey or brown sugar

ANGELICA SYRUP AND COUGH STICKS

This recipe makes both syrup and cough sweets. Sip the syrup to ease coughs and chew one of the sweets every few hours to ease a sore throat.

Bring 1 cup of water and 1 cup of sugar to the boil, stirring until the sugar dissolves. Add the angelica stems and cover. Simmer on low heat until they are soft, for about an hour. Strain and bottle the syrup. Reserve the stems.

Mix the remaining water and sugar together and stir over a low heat until the sugar is dissolved. Add the angelica stems and simmer for 1 hour. Strain and add to the syrup that you have already bottled. Dry the stems on a rack with a tray underneath to catch the drips. Sprinkle them with castor sugar and seal in an airtight container.

- 2 cups water
- 2 cups sugar
- 1 cup angelica stems, cut into 3 cm pieces
- castor sugar

ELDERBERRY AND SAGE COUGH SYRUP

This is an extremely soothing cough syrup and it is full of vitamin C. It can also be added to herbal tea to sweeten it.

Place the sage, thyme and cloves in a pot and add the water. Simmer for 10 minutes. Strip the berries off their stalks and add to the pot. Add the sugar and slowly bring to a simmer, making sure all the sugar dissolves. Cover and simmer for about 15 minutes. Remove from the heat, strain through a fine sieve, mashing the berries to get as much juice out as possible, and bottle.

- ½ cup sage and thyme leaves
- 5 cloves
- 1½ cups water
- about 5 bunches elderberries
- 1 cup sugar

PLUM BLOSSOM VINEGAR

This was one of the first herbal remedies I ever made. It was from one of Margaret Roberts' books and she recommends its use as a gargle for sore throats. I have found it to be a wonderful recipe for people who use their voices — singers, public speakers, lecturers and the like. As a gargle it soothes and strengthens stressed throats and voice boxes. It is the ideal remedy for a sore throat after a night out shouting over loud music!

Mix 1 cup of blossoms with the vinegar in a bottle and leave in the sun for 10 days, shaking regularly. Strain and add the second cup of blossoms. Leave to infuse for 6 weeks. Strain and pour into a sterilised bottle and seal.

To use, dilute in some water and gargle.

- 1 cup plum blossoms
- 3 cups apple cider vinegar
- 1 additional cup plum blossoms

FREQUENT FLYER BALM

One of the quickest ways to catch a cold is to catch an aeroplane. Studies show that we are 100 times more likely to become infected on a plane than on the ground. One of the main reasons for this is the dry cabin air. Our first line of defence against the cold virus is the mucus in our nose. In the low humidity of the cabin, this instantly dries up, leaving us vulnerable. If you are a frequent-flyer bug catcher, try using this balm in and around your nostrils before and during the flight.

Mix the herbs and almond oil together in a double boiler or a bowl set over water and heat it gently for about 3 hours. (The temperature should not go over 100 °C.) Strain the mixture into a bowl or jug. Wipe the double boiler with paper towel and heat the beeswax until it is just melted. Remove from the heat and add the rosehip and rose essential oil. Decant into a sterilised jar.

- 1 tablespoon borage leaves and flowers
- 2 tablespoons rosemary leaves and flowers
- 2 tablespoons bulbine leaves
- 2 tablespoons calendula leaves and flowers
- ⅓ cup almond oil
- 15 g beeswax, grated or chopped
- 1 teaspoon rosehip oil
- 5 drops rose essential oil

IMMUNE-BOOSTING TONIC

Start taking this delicious pick-me-up tonic from late autumn throughout winter to prevent colds, coughs and flu from even thinking about starting.

- 2–3 cinnamon sticks
- 2 tablespoons fresh ginger, chopped
- 1 tablespoon dried echinacea root
- 3 cups water
- 2 tablespoons each fresh nettle, anise hyssop, thyme and heartsease leaves and flowers
- honey
- 10 tablespoons fresh elderberries

Place the cinnamon sticks, ginger and echinacea root in a pot and add the water. Bring to the boil and simmer, covered, for 5 minutes. Add the fresh herbs, cover and remove from the heat. Leave until cooled and strain. For each cup of infusion add 1 ½ cups of honey. Stir in the elderberries and heat. Boil for 6–8 minutes, until it has thickened slightly and the elderberries have softened. Mash the elderberries with a wooden spoon. Remove from the heat and pour into a sterilised bottle. Drink a teaspoon or two every morning.

T & T TONSIL SPRAY

The first sign of a throat infection coming on is usually swelling and we feel this by becoming aware of our swallowing – we normally don't even notice when or how often we swallow. Start using this thyme and tea tree throat spray the minute you start feeling any 'throat awareness' and it will help kill off the infection.

Spread the thyme out in a warm place to dry for a day to reduce its moisture content. Place in a sterilised bottle and pour in the vinegar. Seal and leave to infuse in a dark spot for 6 weeks, shaking every now and then. Strain the vinegar. Mix the salt and distilled water together in a spray bottle. Add the essential oils and infused vinegar and mix well. Store in a cool dark place. To use, shake well and spray the mixture on the back of your throat and tongue.

- ¼ cup fresh thyme
- 1 cup apple cider vinegar
- ½ teaspoon sea salt
- 1 cup distilled water
- 10 drops each tea tree, peppermint and eucalyptus essential oils

Aches, pains, sprains and bruises

ACHY PAINY ANTI-INFLAMMATORY TEA

This warming and anti-inflammatory tea will help ease sciatica, sore arthritic joints and any other inflammatory condition such as shingles.

Crush the cardamom, cloves and ginger in a pestle and mortar until powdered (or grind in a spice grinder). Add all the other ingredients and mix together well. Store in an airtight container. To drink, combine 1 tablespoon of the mixture with 1 cup of milk and 1 cup of water. Bring to the boil and simmer for 20 minutes. Strain, add sugar to taste and whisk until frothy.

- ¼ teaspoon cardamom seeds
- ¼ teaspoon cloves
- 3 tablespoons dried ginger
- 3 tablespoons dried turmeric
- pinch of black pepper
- pinch of cayenne pepper
- 1 teaspoon cinnamon
- 1 cup mixed dried alfalfa, rosemary, nettle and oregano leaves, crushed
- 1 tablespoon black tea leaves

MASSAGE SALVE FOR ACHES AND PAINS

This helps ease painful rheumatism and arthritis. It includes rosemary to increase circulation and cloves to help ease pain. Be warned, though, the turmeric does stain.

- 1 tablespoon each fresh rosemary leaves and flowers, fresh chickweed leaves and flowers, fresh or dried gotu kola leaves, fresh rue leaves and stems
- 180 ml vegetable oil
- 2 tablespoons ground cloves
- ¼ teaspoon ground mustard seeds
- 2 tablespoons ground coriander seeds
- ¼ cup turmeric powder
- 180 ml vodka
- 10 drops each chamomile and St John's wort essential oils
- ¼ teaspoon borax powder

Combine the herbs and vegetable oil and heat over a very low heat for 3 hours. (The temperature should not go over 100 °C.) Remove from the heat and strain. Place the cloves, mustard and coriander seeds, turmeric and vodka in a pan and heat very gently for 5–8 minutes. Don't let it simmer otherwise you will lose the oils. Strain and leave to cool. Add equal proportions of infused oil and alcohol to a sterilised bottle. Add the essential oils and borax, shake well and seal.

HEADACHE BALM

Rub this balm on your temples or the back of your neck to ease a headache. It works especially well for a tension headache.

Mix the herbs, seeds and oil together in a double boiler or a bowl set over water and heat it gently for about 3 hours. (The temperature should not go over 100 °C.) Strain the mixture into a bowl or jug. Wipe the double boiler with paper towel and heat the beeswax until it is just melted. Add the infused oil and stir until it is well mixed. Remove from the heat and stir often as it cools. Transfer to a sterilised storage jar.

- 1 tablespoon chopped bay leaves
- 1 tablespoon coriander seeds, crushed
- 2 tablespoons lavender flowers
- 2 tablespoons mint leaves
- 2 tablespoons rosemary leaves and flowers
- 2 tablespoons tansy leaves
- ⅓ cup almond oil
- 15 g beeswax, grated or chopped

HOT DIGGEDY JOINT SALVE

Containing cayenne pepper, ginger, horseradish and mustard, this salve is a deep heat treatment for sore joints. It increases circulation and blood flow, promoting healing and reducing pain. Just remember to wash your hands after applying it.

- ½ cup grated fresh horseradish root
- 3 cm ginger, grated
- ¾ cup vegetable oil
- 2 tablespoons black mustard seeds, crushed
- 10 bay leaves
- ½ cup fresh oregano leaves
- 180 ml vodka
- ¼ teaspoon cayenne pepper
- ¼ teaspoon borax powder

Combine the horseradish, ginger and vegetable oil and heat over a very low heat for 3 hours. (The temperature should not go over 100°C.) Remove from the heat and strain. Place the mustard seeds, bay leaves and oregano and vodka in a pan and heat very gently for 5–8 minutes. Don't let it simmer otherwise you will lose the oils. Strain and leave to cool.

Add equal proportions of infused oil and alcohol to a sterilised bottle. Add the cayenne pepper and borax, and seal. Shake well before using.

NEURALGIA BREW

The pain from neuralgia can be unbearable, leading to depression and insomnia. This tea uses herbs that contain antispasmodic and anti-inflammatory ingredients to soothe the nervous system, plus cleansing burdock to speed up healing. Combined with the anti-inflammatory and purifying properties of celery, these ingredients make up a cleansing, painkilling and soothing tea. It can also be used by shingles sufferers.

- 1 tablespoon dried burdock root
- 1 tablespoon coriander seeds, crushed
- 1 tablespoon celery seeds, crushed
- 3 cups water
- 1 tablespoon dried California poppy leaves and flowers
- 1 tablespoon each dried chamomile and lavender flowers
- 1 tablespoon dried gotu kola leaves
- 1 tablespoon dried St John's wort
- ¾ cup just-boiled water
- 5 celery sticks, juiced

Place the burdock root, coriander seeds and celery seeds in a pan with the water. Bring to the boil and reduce the heat to a simmer. Cover and leave for 1 hour. Leave to cool and strain. While the decoction cools, place the remaining herbs in a jug and cover with the just-boiled water. Leave to steep for 15 minutes and strain. Mix the decoction with the infusion and the celery juice and drink 1 cup three times a day.

SHINGLES SOOTHER

- 150 ml vegetable oil
- 2 tablespoons coriander seeds, crushed
- ¼ cup peppermint leaves
- ¼ cup St John's wort flowers
- 1 cayenne pepper, sliced
- 3 cm piece of fresh ginger, chopped
- ¼ cup rue leaves and stems
- 2 tablespoons turmeric
- 30 g beeswax, grated or chopped

Turmeric is an effective anti-inflammatory that can be used internally or externally. This painkilling balm is soothing and anti-inflammatory, and increases circulation. Apply it as soon as you feel the first warning signs of shingles. It can also be used by neuralgia sufferers, but be warned: the turmeric can stain both skin and clothing. Be sure to wash your hands after applying the mixture.

Mix the oil, seeds, herbs and turmeric together in a double boiler or a bowl set over water and heat it gently for about 3 hours. (The temperature should not go over 100 °C.) Strain the mixture into a bowl or jug. Wipe the double boiler with paper towel and heat the beeswax until it is just melted. Add the infused oil slowly, stirring all the time until it is mixed. Remove from the heat and transfer to a jar.

Digestion, the stomach and urinary system

INDIGESTION TEA

This mixture aids digestion and soothes the system.

- 2 tablespoons each fennel, dill, coriander and caraway seeds, crushed
- 2 tablespoons dried mint leaves
- 1 tablespoon mustard seeds
- 2 tablespoons dried chamomile flowers

Mix all the ingredients together and store in an airtight container. Add 1 teaspoon of the mixture to 1 cup of boiling water and steep for 10 minutes. Strain and sweeten with honey.

TENSE TUMMY TINCTURE

This tincture calms a nervous stomach and aids digestion. It also helps expel wind.

Mix the herbs together in a large bottle and fill with the vodka, shaking to remove any air bubbles. Seal and leave in a dark spot for two weeks. Strain the herbs out and pour the liquid back into the bottle.

- 2 tablespoons chopped fresh peppermint
- 1 tablespoon chopped fresh thyme
- 2 tablespoons dried chamomile flowers
- 1 tablespoon chopped fresh yarrow
- 1 tablespoon chopped dried liquorice root
- good-quality vodka

AFTER-DINNER MINT LIQUEUR

Serve this great-tasting liqueur at the end of a dinner party. This tummy-soothing digestive will also help relieve wind, heartburn and colic.

Mix the herbs together in a large bottle and fill with the vodka, shaking to remove any air bubbles. Seal and leave in a dark spot for two weeks. Strain the herbs out and pour the liquid back into the bottle. Add honey to taste. Try using port or rum instead of vodka for a different flavour.

- ½ cup chopped fresh angelica root (or ¼ cup dried)
- ¼ cup chopped fresh peppermint
- ¼ cup anise hyssop leaves
- 1 tablespoon crushed fennel seeds
- 2 tablespoons dried chamomile flowers
- good-quality vodka
- honey

HAIR-OF-THE-DOG DETOXIFYING TINCTURE

Mix up a batch of this tincture and give it as Christmas presents. This mix will cleanse your liver, blood, colon and kidneys. It's especially good to detoxify the body after a bit too much partying.

Place the herbs in a large bottle and cover with the alcohol. Seal and leave in a cool place for two weeks. Give it a shake every few days. Strain the tincture into a bowl. Wrap the herbs in cheesecloth and twist it tightly to squeeze the last of the tincture out. Use a funnel to pour into a dark glass bottle and seal.

- ¼ cup dried alfalfa leaves and flowers
- ¼ cup dried milk thistle leaves and seeds
- 2 tablespoons each clover, nettles and selfheal leaves
- 3 tablespoons flax seed, ground
- 2 tablespoons each chopped burdock and dandelion roots
- 1 teaspoon ground licorice root
- 1 litre good-quality vodka, port or rum

TRAVEL-SICKNESS SWEETS

When I was a child I suffered from car sickness. I remember Mom feeding me crystallised ginger sweets to help settle my stomach and the smell of ginger takes me back to car journeys heading on holiday. These sweets also help ease morning sickness.

- 2 cups fresh ginger, peeled and chopped
- ¼ teaspoon cardamom seeds, crushed
- 3 tablespoons fresh peppermint, chopped
- sugar
- pinch of salt
- icing sugar

Place the ginger and cardamom seeds in a pot and cover with water. Bring to the boil and simmer for 30–40 minutes. Strain and reserve the water and ginger separately. Place the mint in a bowl and pour the hot ginger water over. Cover and leave to steep for 15 minutes. Strain and discard the mint. Pour the ginger and mint infusion into a pot, measuring as you do so. For every 1 cup of ginger infusion add 1 cup of sugar. Heat the mixture, stirring until the sugar has dissolved. Add the salt and bring to a rolling boil. Cook, stirring occasionally, until it solidifies when dropped into cold water. Stir in the chopped ginger and remove from the heat. Pour into a greased pan or sweet moulds. Score crisscross lines while it is still soft and once the sweets are set break into pieces and dust with icing sugar. Seal in an airtight container.

HORSERADISH AND CRANBERRY BLADDER CLEANSER

Cystitis, or inflammation of the bladder, is a common and painful ailment that affects women more than men. Cranberries help prevent and clear cystitis by increasing the acidity of the urine so bacteria can't develop. Goldenrod's antiseptic and diuretic properties help cleanse the system. Horseradish contains mustard oil that helps clear any infection.

Place the horseradish and goldenrod in a bowl and pour the water over. Cover and leave to infuse for 10–15 minutes. Strain and leave to cool. Mix with the cranberry juice and drink 1 cup twice a day.

- 1 teaspoon fresh horseradish root, grated
- 1 teaspoon goldenrod flowers and leaves
- 1 cup just-boiled water
- 1 cup cranberry juice

Tension, stress, mind and emotion

SLEEP-EASY TEA

A deep night's sleep is one of the most healing things we can give ourselves. If you suffer regularly from restless sleep or sleeplessness, it could impact on your long-term health. Drink a small cup of this tea before going to bed to ensure a good night's rest.

Mix all the herbs together and store in an airtight container. To drink, infuse ½ teaspoon in ¼ cup of just-boiled water for 10 minutes. Strain and mix with ¼ cup of warm milk. Stir in the nutmeg.

- 2 tablespoons each dried, chamomile and lavender flowers, California poppy root, catnip leaves and St John's wort and anise hyssop leaves and flowers
- ¼ cup just-boiled water
- ¼ cup warm milk
- pinch of nutmeg

DREAM ENHANCER

Place a sachet under your pillow or burn these herbs as incense to encourage lucid dreams and help you remember your dreams.

Mix all the herbs together. Either gather them up in a sachet to put under your pillow or place them in an oil burner and burn before sleeping.

- 2 tablespoons each dried sage, bay leaves, African wormwood, mugwort, motherwort and catnip

BRAIN-BOOSTING TONIC WINE

This wine contains rosemary and sage to improve memory, gotu kola and black pepper to sharpen concentration and basil to relieve tiredness and give the mind strength and clarity.

- 10 black peppercorns
- ¼ cup fresh rosemary leaves
- ¼ cup fresh gotu kola leaves
- ¼ cup fresh basil leaves
- ¼ cup sage leaves
- 1 bottle of good red wine

Crush the peppercorns and bruise the herbs in a pestle and mortar and then place in the bottle of wine and reseal. Keep it in a dark spot for two weeks, shaking every day. Drink a small cup every evening.

PICK-ME-UP BALM

- ½ cup mixed rose petals, rose-scented pelargonium, lavender, California poppy, selfheal, borage and chamomile
- 1 cup vegetable oil
- 25 g beeswax, grated or chopped
- 10 drops each rose-scented pelargonium, bergamot, lavender and rose essential oils

Carry this balm with you and if you are feeling stressed, tired or irritable, rub a little around your ears and on your neck to calm you down.

Combine the herbs and vegetable oil and heat over a very low heat for 3 hours. (The temperature should not go over 100 °C.) Remove from the heat and strain. Wipe the double boiler with paper towel and heat the beeswax until it is just melted. Add the infused oil and stir until it is well mixed. Remove from the heat, add the essential oils and stir often as it cools. Transfer to a sterilised storage jar.

DR FEELGOOD TEA

If you are feeling a bit blue, try this tea to make you feel more cheerful. And don't forget: one of the best remedies for depression is to get out and do some exercise.

Mix the herbs together and seal in an airtight container. To drink, infuse 2 teaspoons of the herb mixture in a mug of just-boiled water for 10 minutes. Strain and add lemon and honey to taste.

- ¼ cup each dried St John's wort, anise hyssop, borage and lemon balm leaves and flowers, and rose-scented pelargonium leaves
- just-boiled water
- lemon
- honey

Heart, blood and circulation

HIBISCUS HEALTHY BLOOD TEA

Hibiscus flowers not only provide a showy display in our gardens, they have many medicinal properties too. High in vitamin C and antioxidants, hibiscus boosts the immune system and protects against cellular damage. It also very effectively reduces high blood pressure. High blood pressure (or hypertension) is a common ailment with a number of causes. Whatever the reasons, this tea helps improve blood flow and circulation, strengthens blood vessels, regulates blood sugar, reduces cholesterol and contains antioxidants, all of which help reduce the risk of hypertension.

Mix all the ingredients together and store in an airtight container. To drink, infuse 1 teaspoon of the mixture in a cup of just-boiled water. Drink a cup every morning.

- ¼ cup dried hibiscus flowers
- 1 teaspoon ginger, finely chopped and dried
- ½ teaspoon garlic cloves, finely chopped and dried
- ¼ cup dried heartsease flowers
- ¼ cup dried motherwort leaves and flowers
- pinch of stevia powder
- 1 teaspoon cinnamon
- ¼ teaspoon cayenne pepper
- 1 teaspoon ground celery seeds

ANTI-ANAEMIA TEA

Thyme is not only rich in iron, it also contains high levels of vitamin C, which aids in the absorption of iron. The lemon slices add more vitamin C and they taste good.

Mix all the ingredients together and store in an airtight container. To drink, infuse 1 teaspoon of the mixture in a cup of just-boiled water. (You can also use the fresh herbs to make an infusion.)

- 2 cups mixed dried watercress, alfalfa, nettle, raspberry and nasturtium leaves
- 1 tablespoon dried thyme
- 1 tablespoon ground celery seeds
- 1 teaspoon ground cumin seeds
- 2 tablespoons lemon slices, dried and crushed

CELLULITE BODY BUTTER

The first step to clearing the build-up of toxins, fat and waste products that cause bumpy cellulite under our skin, is exercise and a healthy diet. The second step is using a body butter that improves the skin's elasticity, increases circulation and clears toxins. Apply this butter in a circular motion to the affected areas after a hot bath or shower.

- ¼ cup angelica leaves
- ¼ cup rosemary leaves
- ¼ cup gotu kola leaves
- ½ cup almond oil
- 20 g beeswax, grated or chopped
- 1 tablespoon witch hazel
- 100 ml basic cream (page 182)
- ¼ teaspoon borax powder
- 10 drops rosemary essential oil
- 10 drops ginger essential oil
- 30 drops grapefruit essential oil

Mix the herbs and almond oil together in a double boiler or a bowl set over water and heat it gently for about 3 hours. Strain the mixture into a bowl or jug. Wipe the double boiler with paper towel and heat the beeswax until it is just melted. Add the infused oil slowly, stirring all the time until it is mixed. Remove from the heat. Mix the witch hazel, basic cream, borax and essential oils together in a pan, and heat over low heat until the borax is dissolved and the mixture is smooth. Remove from the heat and slowly add the beeswax and oil mixture, stirring all the time, until well mixed. Decant into sterilised jars.

NOSEBLEED BALM

- 3 tablespoons each of yarrow and selfheal leaves
- 1 tablespoon salad burnet root, chopped
- ¼ cup olive oil
- 1 tablespoon almond oil
- 15 g beeswax, grated or chopped

These three herbs have been used for centuries to staunch bleeding. However, if you have a nosebleed, it is not the most comfortable thing to stick fresh leaves up your nostrils. This balm helps stop a bleeding nose and also keeps the membranes lubricated to prevent it from happening again.

Mix the herbs and oils together in a double boiler or a bowl set over water and heat it gently for about 3 hours. Strain the mixture into a bowl or jug. Wipe the double boiler with paper towel and heat the beeswax until it is just melted. Add the infused oil and stir until mixed. Remove from the heat and stir often as it cools. Transfer to a sterilised storage jar.

Women

MENOPAUSE TEA

This combination of herbs will help reduce the mood swings, hot flushes and night sweats that can accompany menopause.

- ¼ cup each of dried evening primrose leaves and flowers, sage, raspberry, parsley and St John's wort leaves
- 3 tablespoons ground flax seed

Mix all the ingredients together and store in an airtight container. To drink, infuse 1 teaspoon of the mixture in a cup of just-boiled water. (You can also use the fresh herbs to make an infusion.)

TIME-OF-THE-MONTH BREW

There are plenty of jokes made about PMS. From: 'They call it PMS because Mad Cow Disease was already taken' to Roseanne Barr's 'Women complain about premenstrual syndrome, but I think of it as the only time of the month that I can be myself.' However, when someone is in the middle of PMS mood swings it is definitely no time for joking. (Unless you want your head bitten off.) This tea helps ease the tension and irritability that accompanies many of us during that time of month.

- ¼ cup dried St John's wort leaves
- ¼ cup each of dried selfheal, yarrow and evening primrose leaves and flowers

Mix all the ingredients together and store in an airtight container. To drink, infuse 1 teaspoon of the mixture in a cup of just-boiled water. (You can also use the fresh herbs to make an infusion.)

Skin conditions

COLD-SORE BALM

If you apply this healing and soothing balm as soon as you feel the warning itchy feeling of a cold sore coming on, it will help diminish the symptoms. Reapply regularly until the cold sore has gone. Repeated treatment will reduce the frequency of cold sores breaking out.

- ½ cup olive oil
- 2 tablespoons almond oil
- ¼ cup mixed calendula petals, peppermint leaves, lavender flowers, lemon balm leaves and St John's wort leaves
- 15 g beeswax, grated or chopped
- 10 g cocoa butter
- 5 drops mandarin essential oil
- 1 teaspoon honey

Mix the olive and almond oils and herbs together in a double boiler or a bowl set over water and heat it gently for about 3 hours. Strain the mixture into a bowl or jug. Wipe the double boiler with paper towel and heat the beeswax and cocoa butter until they are just melted. Add the infused oil and stir until it is well mixed. Remove from the heat, mix in the essential oil and honey and stir often as it cools. Transfer to a sterilised storage jar.

GREEN GOODNESS OINTMENT

This is a fabulous all-purpose ointment to treat insect bites and stings, minor wounds, itchy skin and rashes. If you don't have all the herbs, adapt it using the ones you do.

- ¼ cup each elderflowers, comfrey leaves, calendula petals, basil leaves, mint leaves and lavender flowers
- 1⅓ cup vegetable oil
- 50 g beeswax, grated or chopped
- 10 g shea butter
- 3 teaspoons aloe vera gel
- few drops each chamomile, lavender and tea tree essential oils

Mix the herbs and vegetable oil together in a double boiler or a bowl set over water and heat gently for about 3 hours. Strain the mixture into a bowl or jug. Wipe the double boiler with paper towel and heat the beeswax until it is just melted. Slowly add the infused oil, stirring until it is completely mixed. Add the shea butter and aloe vera gel and stir until mixed in. Remove from the heat, mix in the essential oils and stir often as it cools. Transfer to a sterilised storage jar.

SUNBURN SOOTHER

This cooling, healing gel is perfect after a day at the beach. It can also be dabbed onto minor burns and insect stings.

- 1 cup mixed lavender and calendula flowers
- 1 cup mixed comfrey and peppermint leaves
- vegetable oil
- 2 tablespoons aloe vera gel

Place the herbs in a sterilised bottle, cover with vegetable oil and leave to infuse for 2 months. Strain, and stir in the aloe vera gel. Store in a dark glass container in the fridge. Rub gently onto sunburned skin. This mixture will keep for up to a month.

SKIN-SOOTHING CREAM

- ½ cup mixed chamomile flowers, borage leaves and flowers, gotu kola leaves, calendula leaves and flowers and comfrey leaves, flowers and roots
- ¾ cup almond oil
- 35 g beeswax, grated or chopped
- ¾ cup basic cream (page 182)
- 3 teaspoons aloe vera gel
- ⅛ teaspoon borax
- 10 drops each rosemary, lavender and rose-scented pelargonium essential oils

This anti-inflammatory and soothing cream is excellent for rashes, dry itchy skin and eczema.

Mix the herbs and oil together in a double boiler or a bowl set over water and heat it gently for about 3 hours. Strain the mixture into a bowl or jug. Wipe the double boiler with paper towel and heat the beeswax until it is just melted. Add the infused oil slowly, stirring all the time until mixed and remove from the heat. Mix the basic cream, aloe vera gel and borax together in a pan and heat over low heat until the borax is dissolved and the mixture is smooth. Slowly add the beeswax and oil mixture, stirring all the time, until well mixed. Stir in the essential oils. Decant into sterilised jars.

HONEY-AND-HERB WOUND HEALER

- 1 cup honey
- ½ cup each dried lavender flowers and calendula petals
- 5 drops lavender essential oil

This combination of honey and herbs is an easy-to-make yet powerful salve to treat cuts and wounds.

Mix the honey with the herbs in a sterilised bottle. Leave it to infuse for about 6 weeks. Mix in the lavender oil. To use, apply the salve directly to the cleaned wound and cover with a bandage. Repeat daily until the wound has healed.

Just For Pets *Recipes to keep them happy and healthy*

You will notice there are many more recipes here for dogs than for cats. In fact there is only one cat recipe here – for a catnip mouse. This is not because I have anything against cats (I share my home with a couple). Cats and dogs are both carnivores, however dogs are more omnivorous than cats and can digest a far wider range of plants, hence are more suitable candidates for herbal medicines. More caution is needed when giving herbs to cats, as many of them are potentially toxic and can even result in death.

HERBAL FLEA POWDER

Use this flea powder to sprinkle onto bedding and apply to dogs' coats to repel fleas. To make a homemade shaker, tie a disc of netting over the top of a bottle.

- 1 tablespoon each dried elder, rosemary, African wormwood, tansy and fennel leaves
- ½ cup cornflour

Crush the leaves in a pestle and mortar until they are powdered. Mix with the cornflour. Keep in an airtight container.

ARTHRITIC RELIEF FOR OLDER DOGS

- ¼ cup each dried alfalfa and clover
- 2 tablespoons dried parsley root, powdered
- 2 tablespoons powdered turmeric

If you notice stiffness in your dogs' joints after they have been sleeping or if they begin to limp after a walk, it is time to start giving them something to ease joint pain. Turmeric is a powerful anti-inflammatory, which works quickly to reduce painful joint swelling. It is accompanied by the strengthening and cleansing qualities of alfalfa, parsley and clover.

Mix all the ingredients together and store in an airtight container. Give your dogs half a teaspoon morning and evening, mixed into their food. If they don't like it on their food, mix it separately in some tasty meat or chicken broth.

HEALTHY COAT TONIC

This supplement will keep your dog's coat shiny and healthy.

- 1 cup dried nettle leaves
- 2 tablespoons flax seed oil

Mix the ingredients together to form a paste. Mix ½ teaspoon into your dog's food per day.

FLY SPRAY FOR DOGS AND HORSES

During summer flies can torment dogs and horses, especially around their ears and the base of their tails. Use this spray regularly on these areas. For more sensitive areas, spray it onto your hands and rub it onto their coats.

- ½ cup weak tansy infusion
- 1 cup mixed lemon balm and lemon grass infusion
- 1 cup apple cider vinegar
- 5 drops citronella essential oil
- 5 drops lavender essential oil
- 5 drops tea tree oil
- 10 drops glycerine

Caution: Both tea tree oil and tansy can be toxic to dogs if they lick it off in large quantities. Do not exceed the dose below and do not spray excessively on small dogs. Do not use citronella lamp oil; it is poisonous.

Mix all the ingredients together in a spray bottle and shake well.

STRESS AND TENSION EASER

Many dogs become stressed with thunder, a visit to the vet or fireworks. A few drops of this tincture added to their drinking water will help ease the symptoms.

- ⅓ cup each dried chamomile flowers, catnip leaves and St John's wort leaves and flowers
- 1 litre vodka

Place the herbs in a bottle and cover with the vodka. Seal and leave in a cool place for 2 weeks. Give it a shake every few days. Strain into a bowl. Wrap the herb in cheesecloth and twist it tightly to squeeze the last of the tincture out. Use a funnel to pour into a dark glass bottle and seal.

ALL-PURPOSE HEALING BALM

- ½ cup olive oil
- 2 tablespoons almond oil
- 3 tablespoons each calendula and clover leaves
- 3 tablespoons each St John's wort and chamomile flowers
- 15 g beeswax, grated or chopped
- 1 tablespoon aloe vera gel
- 1 tablespoon bulbine gel

Use this healing and soothing balm for minor cuts, wounds, insect bites and rashes on dogs. It is also good for itchy stitches after surgery. Any of these herbs can also be used as a fresh infusion as a wash to reduce itchy rashes.

Mix the oil and herbs together in a double boiler or a bowl set over water and heat gently for about 3 hours. Strain the mixture into a bowl or jug. Wipe the double boiler with paper towel and heat the beeswax until it is just melted. Add the infused oil slowly, stirring all the time until it is mixed. Stir in the aloe vera and bulbine gel. Remove from the heat and stir often as it cools. Transfer to a storage jar and store in the fridge.

CATNIP MOUSE

Top of the list of herbs for cats is of course catnip. Cats love rolling in and rubbing against catnip. This is a great present for a cat who doesn't have catnip in the garden.

Fold the felt square in half and cut out half a heart shape, so when it is opened it creates a full heart shape. Fold it in half again and sew the edges together, leaving a small gap. Stuff the mouse with catnip and sew the gap closed. Sew a bit of rope on for a tail and draw eyes and whiskers on the felt.

- ½ cup dried catnip
- a square of felt

HERBS FOR HENS

Hens will make a beeline for certain herbs because they know they are full of nutrients: alfalfa, chickweed, comfrey, nasturtium and dandelion are all good herbs to add to your hens' food. In the wild, hens will use herbs in their nests to keep parasites and bacteria at bay. Use the following to help keep your hens parasite free.

Leave the herbs to dry for a day, then add them to the nesting boxes and mix them with the hay in the run to prevent parasites.

- 3 cups of any of the following herbs: peppermint, African wormwood, catnip, tansy, oregano, lavender, rosemary, sage, thyme, fennel or feverfew

In The Garden *Recipes for an organic pesticide- and chemical-free garden*

The garden is an area of our home that has been the worst affected by the use of chemicals. Harmful chemical fertilisers deplete our soil and upset the balance of insect, plant and microscopic life. Toxic sprays kill indiscriminately and break down slowly, continuing their damage for many years. Herbal remedies do none of this. They are a safe and non-toxic option to add nutrients and repel harmful insects the natural way.

COMFREY TEA

This is an extremely nutritious feed for hungry vegetables.

- ½ bucket comfrey leaves
- boiling water

Cover the leaves with boiling water and let them rot for about a month, stirring every now and then. Keep it covered as it smells really foul but the plants, especially tomatoes, love it. Dilute the tea 1 part comfrey tea to 10 parts water and use as a soil drench or foliar spray. If you use a bucket with a tap on the side, you can keep topping up the brew and pour it off whenever you need it.

ANTI-FUNGAL SPRAY

This is a good spray to use on powdery mildew and other fungal infections that occur in damp, hot conditions.

Cover the herbs with hot water, stir and leave to stand for 24 hours. Strain out the leaves and stems and place them in another bucket. Repeat with more warm water to make a second infusion. Strain and place the herbs in the compost. To use, mix 4 tablespoons of dish-washing liquid to 1 bucket of infusion. The spray will keep for up to a month. Spray onto affected plants every few days. Make sure you spray underneath the leaves as well as on top.

- 1 bucket mixed chamomile and lavender flowers, clover, horseradish, lemon grass, oregano, thyme, mustard, tea tree, lavender and yarrow leaves
- hot water
- dish-washing liquid

INSECT-REPELLING SPRAY

There are many herbs with insect-repelling properties and it is a good idea to grow a selection of these in amongst your vegetables and herbs. Many of them grow abundantly and are good candidates to be used for natural insecticides. In this recipe I have used elder leaves. You can also use tansy, feverfew, thyme or African wormwood.

Half fill a bucket with elder leaves and stems. Add the just-boiled water, stir and leave to stand overnight. Strain out the leaves and stems and add them to your compost. Add the dish-washing liquid and mix. The spray will keep for up to a month. Spray onto affected plants every few days as the herbal insecticide will break down quickly. Make sure you spray underneath the leaves as well as on top.

- ½ bucket elder leaves and stems
- just-boiled water
- 2 tablespoons dish-washing liquid

A QUICK GUIDE TO HEALING HERBS

SKIN AND HAIR

Uses	Herb	Parts used	Method	Page
Ageing and wrinkles	calendula	leaves and flowers	infused oil in face cream	70
	evening primrose	seeds	cream	100
	gotu kola	leaves and stems	essential and infused oil, infusion internally	111
Boils and abscesses	burdock	roots	decoction internally	68
	chickweed	leaves and flowers	poultice or cream	80
	clover	leaves and flowers	infusion internally	84
	dandelion	leaves	infusion as a wash	92
	echinacea	leaves, flowers and roots	infusion internally, to boost immune system and purify the blood	96
	flax	seeds	poultice	106
	lemon verbena	leaves	ointment	125
	lovage	leaves, seeds and roots	decoction as a wash	126
	nasturtium	seeds	poultice	138
	soapwort	leaves and roots	ointment	157
	thyme	leaves and flowers	infusion internally	168
	watercress	leaves	poultice or tincture internally	171
Burns (and sunburn)	aloe vera	leaves, gel inside leaves	fresh gel	48
	bitter aloe	leaves, gel inside leaves	fresh gel	60
	bulbine	leaves, gel inside leaves	fresh gel	67
	calendula	petals	ointment	70
	cancer bush	leaves	infusion as a wash	73
	chamomile	flowers	ointment	78
	clover	flowers and leaves	ointment	84
	comfrey	flowers, leaves and roots	cream	86
	echinacea	flowers, leaves and roots	decoction as a wash	96
	gotu kola	leaves and stems	ointment	111
	houseleek	leaves, gel inside leaves	fresh gel	118
	lavender	flowers	cream	119
	nettle	leaves and flowers	infusion as a wash	140
	peppermint	leaves and flowers	ointment, oil infusion or essential oil	132
	salad burnet	leaves	infusion as a wash, ointment	154
	St John's wort	leaves and flowers	infusion as a wash, infused oil	160

SKIN AND HAIR

Uses	Herb	Parts used	Method	Page
Cellulite	angelica	leaves	cream, infusion internally	50
Chilblains	chilli	fruit	ointment	82
	elder	flowers	cream	98
	mustard	seeds crushed to powder	foot bath	136
Cold sores	bulbine	leaves	fresh gel	67
	lemon balm	leaves and flowering tips	infusion or essential oil in an ointment	121
	St John's wort	flowers	cream or infused oil	160
Dry and itchy skin	aloe vera	leaves, gel inside leaves	fresh gel added to moisturising cream	48
	bitter aloe	leaves, gel inside leaves	fresh gel added to moisturising cream	60
	borage	leaves and flowers	steam inhalation, infused oil, cream	64
	bulbine	leaves, gel inside leaves	fresh gel	67
	calendula	leaves and flowers	ointment	70
	chamomile	flowers	cream	78
	comfrey	leaves, flowers and roots	cream	86
	elder	flowers	cream	98
	evening primrose	seeds	infused oil	100
	feverfew	leaves and flowers	cream	104
	heartsease	flowers, aerial parts and roots	cream	114
	rosemary	leaves and flowers	infused or essential oil added to cream	148
	soapwort	leaves and roots	infusion as a wash, ointment	157
Eczema	bitter aloe	leaves, gel inside leaves	fresh gel, infusion of leaf as a wash, gel added to moisturising cream	60
	borage	seeds	seed oil	64
	bulbine	leaves, gel inside leaves	fresh gel added to moisturising cream	67
	burdock	root	decoction internally	68
	calendula	leaves and flowers	ointment	70
	chamomile	flowers	ointment	78
	chickweed	leaves and roots	infusion as a wash, cream, infused oil	80
	clover	leaves and flowers	ointment	84
	dandelion	leaves, flowers, stems	infusion internally or externally as a wash, juice	92
	echinacea	flowers, leaves and roots	decoction as a wash	96

SKIN AND HAIR

Uses	Herb	Parts used	Method	Page
Eczema cont.	evening primrose	seeds	infused oil	100
	flax	seeds	oil	106
	goldenrod	leaves and flowers	ointment	110
	gotu kola	leaves and stems	cream	111
	heartsease	flowers, aerial parts and roots	cream	114
	lavender	flowers	essential oil added to cream	119
	nettle	leaves and flowers	infusion internally and as a wash	140
	parsley	leaves	infusion as a wash	142
	pelargonium	leaves and flowers	infused or essential oil added to cream	144
	rosemary	leaves and flowers	infusion as a wash, infused or essential oil added to cream	148
	salad burnet	leaves and roots	ointment	154
	soapwort	leaves and roots	infusion as a wash, ointment	157
Fungal conditions	aloe vera	leaves, gel inside leaves	fresh gel	48
	bitter aloe	leaves, gel inside leaves	fresh gel	60
	burdock	root	decoction as a wash	68
	calendula	leaves and flowers	tincture	70
	cancer bush	leaves	infusion as a wash	73
	chamomile	flowers	ointment	78
	clover	flowers and leaves	ointment	84
	coriander	seeds and leaves	cream	88
	goldenrod	leaves and flowers	ointment	110
	lavender	flowers	essential oil added to cream	119
	lemon grass	stems and leaves	essential oil added to ointment	123
	marjoram and oregano	leaves and flowers	cream	128
	mustard	seeds	infused oil added to cream	136
	rue	leaves and stems	ointment	150
	tea tree	leaves and branches	infusion as a wash, infused oil or essential oil diluted	166
	thyme	leaves and flowers	infusion as a wash	168
	watercress	leaves	infusion internally or as a wash, juice, poultice, ointment	171

SKIN AND HAIR

Uses	Herb	Parts used	Method	Page
Hair	alfalfa	leaves and flowers	infusion or tincture internally, for split ends or thinning hair	46
	gotu kola	leaves and stems	infusion or tincture internally, to stimulate hair growth	111
	nettle	root	decotion externally, to stimulate hair growth and treat dandruff	140
	parsley	leaves and seeds	infusion (leaves) internally, to strengthen hair, externally (seeds) to remove lice	142
	rosemary	leaves and flowers	infusion as a hair rinse for shiny hair, infused oil to condition hair	148
	watercress	leaves	infusion externally to prevent hair loss	171
Itchy, irritated or infected skin	aloe vera	leaves, gel inside leaves	fresh gel	48
	angelica	leaves	cream	50
	bergamot	leaves and flowers	infusion externally or poultice	58
	bitter aloe	leaves, gel inside leaves	fresh gel	60
	borage	leaves and flowers	infused oil, cream	64
	bulbine	leaves, gel inside leaves	fresh gel or added to moisturising cream	67
	calendula	leaves and flowers	ointment	70
	cancer bush	leaves	infusion internally	73
	chamomile	flowers	ointment, essential oil or infused oil	78
	chickweed	leaves and flowers	infused oil, cream	80
	clover	flowers and leaves	ointment	84
	coriander	seeds and leaves	cream, infused oil	88
	elder	flowers	cream	98
	evening primrose	seeds	infused oil	100
	feverfew	leaves	cream	104
	flax	seeds	oil	106
	goldenrod	leaves and flowers	ointment	110
	heartsease	flowers, aerial parts and roots	cream	114
	marjoram and oregano	leaves and flowers	cream, essential or infused oil	128
	nettle	flowers, leaves and roots	tincture or infusion as a wash	140
	parsley	leaves	infusion as a wash	142
	pelargonium	*P. graveolens* leaves	infused or essential oil added to cream	144

SKIN AND HAIR

Uses	Herb	Parts used	Method	Page
Itchy, irritated or infected skin cont.	peppermint	leaves and flowers	infusion or tincture as a wash, ointment, infused oil or essential oil	132
	raspberry	leaves, fruit and stems	tincture as wash	62
	salad burnet	leaves and roots	ointment	154
	soapwort	leaves and roots	infusion as a wash and ointment	157
	watercress	leaves	infusion (internally or as a wash), juice, poultice, ointment	171
Pimples and acne	burdock	root	decoction internally and as a wash	68
	calendula	leaves and flowers	infusion as a wash, tincture externally	70
	dandelion	leaves	infusion internally and as a wash (leaves flowers, stem), juice (leaves)	92
	echinacea	flowers, leaves and roots	decoction as a wash	96
	evening primrose	seeds	oil	100
	goldenrod	leaves and flowers	ointment	110
	lemon verbena	leaves	ointment	125
	lovage	leaves, seeds and roots	decoction as a wash	126
	parsley	leaves	infusion internally or as a wash	142
	pelargonium	*P. graveolens* leaves	infusion as a wash, essential oil added to cleanser	144
	rosemary	leaves and flowers	infusion as a wash, infused or essential oil added to cream	148
	soapwort	leaves and roots	infusion as a wash	157
	sorrel	leaves	infusion as a wash, ointment	158
	tea tree	leaves and branches	infusion as a wash, infused oil or essential oil diluted	166
	watercress	leaves	infusion (internally or as a wash), juice, poultice, ointment	171
Psoriasis	burdock	roots	decoction internally and as a wash	68
	chickweed	leaves and flowers	cream, infused oil, infusion as a wash	80
	clover	leaves and flowers	ointment	84
	evening primrose	seeds	infused oil	100
	flax	seeds	oil	106
	goldenrod	leaves and flowers	ointment	110
	gotu kola	leaves and stems	cream	111
	parsley	leaves	infusion as a wash	142

SKIN AND HAIR

Uses	Herb	Parts used	Method	Page
Psoriasis cont.	rosemary	leaves and flowers	infusion as a wash, infused or essential oil added to cream	148
	salad burnet	leaves and roots	ointment	154
	soapwort	leaves and roots	infusion as a wash, ointment	157
Rashes	aloe vera	leaves, gel inside leaves	fresh gel	48
	basil	leaves	rub fresh leaf on	55
	bitter aloe	leaves, gel inside leaves	fresh gel	60
	borage	leaves and flowers	infused oil, cream	64
	bulbine	leaves, gel inside leaves	fresh gel, cream	67
	chickweed	leaves and flowers	infused oil	80
	coriander	leaves and seeds	infused oil	88
	evening primrose	seeds	oil	100
	heartsease	flowers, aerial parts and roots	infusion as a wash, poultice	114
	houseleek	leaves, gel inside leaves	fresh gel	118
	nettle	leaves and flowers	infusion as a wash	140
	soapwort	leaves and roots	infusion as a wash, ointment	157
	tea tree	leaves and branches	infusion as a wash	166
	watercress	leaves	infusion (internally or as a wash), juice, poultice, ointment	171
Scabies	tansy	leaves	infusion as a wash	163
Scarring (reduction)	aloe vera	leaves, gel inside leaves	fresh gel	48
	comfrey	flowers, leaves and roots	cream	86
	evening primrose	seeds	oil	100
	gotu kola	leaves and stems	ointment	111
Sunburn and burns	aloe vera	leaves, gel inside leaves	fresh gel	48
	bitter aloe	leaves, gel inside leaves	fresh gel	60
	bulbine	leaves, gel inside leaves	fresh gel, cream	67
	calendula	leaves and flowers	ointment	70
	cancer bush	leaves	infusion as a wash	73
	chamomile	flowers	essential oil or infused oil in a lotion or ointment	78
	clover	flowers and leaves	ointment	84
	comfrey	flowers, leaves and roots	cream	86

SKIN AND HAIR

Uses	Herb	Parts used	Method	Page
Sunburn and burns cont.	echinacea	flowers, leaves and roots	decoction as a wash	96
	gotu kola	leaves and stems	ointment	111
	houseleek	leaves, gel inside leaves	fresh gel	118
	lavender	flowers	essential oil or infused oil in a lotion	119
	nettle	leaves and flowers	infusion as a wash	140
	peppermint	leaves and flowers	ointment, oil infusion or essential oil	132
	salad burnet	leaves and roots	infusion as a wash for sunburn	154
	St John's wort	leaves and flowers	infusion or tincture as a wash, infused oil	160
Varicose veins	calendula	leaves and flowers	ointment, infused oil	70
	comfrey	leaves, flowers and roots	cream	86
	gotu kola	leaves and stems	cream, infusion (internally or externally as a compress)	111
	heartsease	flowers, aerial parts and roots	cream, poultice	114
	rue	leaves and stems	infusion as a wash, ointment	150
	yarrow	flowers and leaves	compress	174
Wounds, cuts, stings and bites (see also Blood vessels and bleeding, page 275)	aloe vera	leaves, gel inside leaves	fresh gel	48
	bergamot	leaves and flowers	infusion externally, poultice	58
	bitter aloe	leaves, gel inside leaves	fresh gel	60
	blackberry and raspberry	leaves, fruit and stems	infusion externally, poultice	62
	bugle	whole plant	ointment	66
	bulbine	leaves, gel inside leaves	fresh gel, cream	67
	calendula	leaves and flowers	ointment, compress	70
	cancer bush	leaves	infusion as a wash	73
	chamomile	flowers	ointment	78
	clover	flowers and leaves	fresh	84
	comfrey	flowers, leaves and roots	cream	86
	echinacea	flowers, leaves and roots	decoction as a wash, tincture externally	96
	evening primrose	seeds	poultice	100
	goldenrod	leaves and flowers	ointment	110
	gotu kola	leaves and stems	ointment	111
	ground ivy	leaves and flowers	infusion as a wash, ointment, compress	113
	houseleek	leaves, gel inside leaves	fresh gel	118
	lavender	flowers	essential oil or infused oil in a lotion	119

QUICK GUIDE

SKIN AND HAIR

Uses	Herb	Parts used	Method	Page
Wounds, cuts, stings and bites cont.	lemon balm	leaves and flowers	ointment	121
	nettle	leaves and flowers	infusion as a wash	140
	oregano	leaves and flowers	cream	128
	parsley	leaves	compress	142
	pelargonium	*P. graveolens* leaves	infusion as a wash	144
	raspberry	leaves	infusion or tincture as a wash	62
	sage	leaves	infusion as a wash	152
	salad burnet	leaves and roots	ointment and poultice	154
	selfheal	leaves	poultice applied to fresh wounds to stop bleeding	156
	St John's wort	flowers	cream or infused oil	160
	tea tree	leaves and branches	infusion as a wash, ointment, infused oil	166
	thyme	leaves and flowers	poultice	168

DIGESTIVE SYSTEM

Uses	Herb	Parts used	Method	Page
Appetite (stimulate and repress)	African wormwood	leaves	infusion or tincture internally, stimulate appetite	44
	aloe vera	gel	infusion internally, stimulate appetite	48
	angelica	leaves	infusion or tincture internally, stimulate appetite	50
	bay tree	leaves	infusion internally, stimulate appetite	56
	cancer bush	leaves	infusion internally, stimulate appetite	73
	chamomile	flowers	infusion internally, stimulate appetite	78
	coriander	leaves and seeds	infusion internally, stimulate appetite	88
	dandelion	roots	tincture internally, stimulate appetite	92
	fennel	seeds	infusion internally, repress appetite	102
	feverfew	leaves and flowers	infusion internally, stimulate appetite	104
	mustard	seeds	infusion internally, stimulate appetite	136
	savory	leaves	infusion or tincture internally, stimulate appetite	155
	sorrel	leaves	infusion internally, stimulate appetite	158
	tarragon	leaves	infusion internally, stimulate appetite	164
Constipation	aloe vera	whole leaf	infusion (gel) or tincture (whole leaf) internally	48
	bitter aloe	whole leaf	infusion or tincture internally	60
	burdock	leaves	infusion internally	68

DIGESTIVE SYSTEM

Uses	Herb	Parts used	Method	Page
Constipation cont.	chicory	roots	syrup	81
	elder	berries	decoction internally	98
	flax	seeds	whole fresh seeds	106
	milk thistle	leaves and seeds	infusion internally	130
	mustard	seeds	infusion internally	136
	peppermint	leaves and flowers	infusion or tincture internally	132
Detoxifying	African potato	corms	infusion or decoction internally	42
	alfalfa	leaves and flowers	infusion or tincture internally	46
	angelica	leaves	infusion internally	50
	bitter aloe	gel	infusion internally	60
	borage	leaves and flowers	infusion internally	64
	burdock	roots	tincture internally	68
	cancer bush	leaves	infusion internally	73
	caraway	seeds	infusion internally	76
	chickweed	leaves and flowers	infusion internally	80
	clover	flowers and leaves	infusion or tincture internally	84
	coriander	seeds	infusion internally	88
	dandelion	leaves, flowers and stems	infusion (leaves, flowers, stem), juice (leaves), decoction (roots) or tincture (roots) internally	92
	fennel	seeds and roots	infusion (seeds) or decoction internally (seeds and root)	102
	feverfew	leaves and flowers	infusion or tincture internally	104
	ginger	rhizomes	decoction or tincture internally	108
	gotu kola	leaves and stems	infusion or tincture internally	111
	ground ivy	leaves and flowers	infusion internally	113
	milk thistle	leaves and seeds	infusion internally	130
	nettle	leaves and flowers	infusion internally	140
	rocket	leaves and seeds	infusion internally	146
	watercress	leaves	infusion internally	171
Diarrhoea and gastro-enteritis	blackberry	leaves	infusion internally	62
	lemon verbena	leaves	infusion or tincture internally	125
	pelargonium	*P. graveolens* leaves	infusion internally	144
	raspberry	leaves	infusion internally	62
	thyme	leaves and flowers	tincture internally	168
	watercress	leaves	infusion internally	171

DIGESTIVE SYSTEM

Uses	Herb	Parts used	Method	Page
Digestion	African wormwood	leaves	infusion or tincture internally, improves slow digestion	44
	alfalfa	leaves and flowers	infusion internally, improves digestion, reduces acidity	46
	aloe vera	gel	infusion internally, improves slow digestion	48
	angelica	leaves	infusion or tincture internally, eases indigestion	50
	bay tree	leaves	infusion internally, to aid digestion	56
	bergamot	leaves and flowers	infusion internally, to relieve indigestion	58
	burdock	leaves and roots	tincture to stimulate digestion (roots), infusion to ease indigestion and stimulate digestion (leaves)	68
	calendula	petals and leaves	infusion to ease digestive upsets	70
	caraway	seeds	infusion internally, to aid digestion	76
	catnip	leaves	infusion internally, to treat tense indigestion	77
	chamomile	flowers	infusion internally, to treat tense indigestion	78
	chickweed	leaves and flowers	infusion internally, to aid digestion	80
	chilli	fruit	infusion or tincture internally, to stimulate digestion	82
	chives	leaves and flowers	infusion internally, to ease indigestion	83
	coriander	leaves and seeds	infusion internally, to aid digestion	88
	dandelion	root	tincture internally, to aid digestion	92
	dill	leaves and seeds	infusion or tincture internally, to aid digestion	94
	fennel	seeds	infusion internally, for indigestion	102
	feverfew	leaves and flowers	infusion internally, to stimulate digestion	104
	flax	seeds	fresh seeds or infusion, to soothe digestive system	106
	ground ivy	leaves and flowers	infusion internally, to soothe indigestion	113
	horseradish	root	infusion internally, to stimulate digestion	116
	lavender	flowers	infusion internally, especially for indigestion due to tension	119
	lemon grass	stems and leaves	infusion or tincture internally	123
	lemon verbena	leaves	infusion or tincture internally, to reduce indigestion and aid digestion	125
	lovage	leaves and seeds	infusion internally, to aid digestion	126
	marjoram and oregano	leaves and flowers	infusion internally, for indigestion	128
	milk thistle	leaves, ground seeds and flowers	infusion or tincture internally, to treat digestive problems	130
	mugwort	leaves and roots	infusion (leaves) or decoction (roots) internally	135

DIGESTIVE SYSTEM

Uses	Herb	Parts used	Method	Page
Digestion cont.	mustard	seeds	infusion internally, aids digestion, speeds up metabolism	136
	parsley	roots	decoction internally, for indigestion	142
	peppermint	leaves and flowers	infusion or tincture internally, for indigestion	132
	raspberry	leaves	infusion internally, to treat digestive disorders	62
	rocket	leaves and seeds	infusion internally, to stimulate digestion	146
	rosemary	leaves	infusion internally, to stimulate digestion	148
	rue	leaves and stems	infusion internally, stimulates digestive system	150
	sage	leaves	infusion internally, especially after fatty meals	152
	savory	leaves	infusion or tincture internally, especially with beans	155
	sorrel	leaves	infusion internally, to stimulate digestion	158
	stevia	leaves	infusion internally, for indigestion	162
	tarragon	leaves	infusion internally, particularly aids digestion of proteins	164
	Vietnamese coriander	leaves	infusion internally, to ease indigestion	170
	watercress	leaves	infusion internally, improves sluggish digestion	171
Flatulence (wind)	angelica	leaves	infusion or tincture internally	50
	caraway	seeds	infusion internally	76
	dill	leaves and seeds	infusion or tincture internally	94
	fennel	seeds	infusion internally	102
	lemon verbena	leaves	infusion or tincture internally	125
	lovage	leaves and seeds	infusion internally	126
	oregano	leaves and flowers	infusion internally	128
	peppermint	leaves and flowers	infusion or tincture internally	132
	rosemary	leaves and flowers	infusion internally	148
	savory	leaves	infusion internally	155
	Vietnamese coriander	leaves	infusion internally	170
Gall bladder	dandelion	leaves, flowers and stems	infusion internally (leaves, flowers, stem), juice (leaves)	92
	milk thistle	leaves, ground seeds and flowers	tincture internally	130
	parsley	leaves, seeds and roots	infusion internally	142

QUICK GUIDE **270**

DIGESTIVE SYSTEM

Uses	Herb	Parts used	Method	Page
Haemorrhoids	elder	leaves	oil infusion with flax oil	98
Irritable bowel syndrome	chamomile	flowers	infusion internally	78
	peppermint	leaves and flowers	infusion or tincture internally	132
Liver problems	aloe vera	gel	infusion or tincture internally, to cleanse liver	48
	angelica	roots	tincture internally, to stimulate liver	50
	cancer bush	leaves	infusion internally	73
	cape gooseberry	fruit	infusion internally	74
	chicory	leaves and flowers	infusion internally	81
	clover	leaves and flowers	infusion or tincture internally, to cleanse liver	84
	coriander	seeds	infusion internally, to cleanse liver	88
	dandelion	leaves, flowers, roots and stems	infusion (leaves, flowers, stem), juice (leaves), tincture (root) or decoction (root) internally, to cleanse and stimulate liver	92 104
	feverfew	leaves and flowers	weak infusion internally, to stimulate liver	116
	horseradish	roots	infusion internally	126
	lovage	leaves and seeds	infusion internally, to cleanse liver	130
	milk thistle	leaves, ground seeds and flowers	tincture internally	156
	selfheal	flowers	decoction internally, to stimulate liver	171
	watercress	leaves	infusion internally, to tone the liver	
Nausea and vomiting	basil	leaves and flowers	infusion internally, essential oil added to inhaler, tincture internally	54
	bergamot	leaves and flowers	infusion internally	58
	chamomile	flowers	infusion internally	78
	feverfew	leaves and flowers	weak infusion internally	104
	ginger	rhizome	decoction or tincture internally	108
	peppermint	leaves and flowers	infusion internally, essential oil added to inhaler	132
Stomach cramps and colic	African wormwood	leaves	infusion or tincture internally	44
	angelica	leaves	infusion or tincture internally	50
	bay tree	leaves	infusion internally	56

DIGESTIVE SYSTEM

Uses	Herb	Parts used	Method	Page
Stomach cramps and colic cont.	calendula	petals and leaves	infusion internally	70
	caraway	seeds	infusion internally	76
	catnip	leaves	infusion internally	77
	chamomile	flowers	infusion internally	78
	dill	leaves and seeds	infusion (seeds) internally for babies, infusion or tincture (leaves and seeds) internally	94
	fennel	seeds	infusion or decoction internally	102
	lavender	flowers	infusion internally, especially for relief of conditions due to tension	119
	lemon balm	leaves and flowers	infusion or tincture internally, especially for relief of conditions due to tension	121
	lemon verbena	leaves	infusion or tincture internally	125
	lovage	leaves and seeds	infusion internally	126
	parsley	root	decoction internally	142
	peppermint	leaves and flowers	infusion or tincture internally	132
	savory	leaves	infusion or tincture internally	155
	Vietnamese coriander	leaves	infusion internally	170
Stomach ulcers	aloe vera	gel	infusion internally	48

MUSCLES, BONES AND JOINTS

Uses	Herb	Parts used	Method	Page
Arthritis and rheumatism	African potato	corms	infusion or decoction internally	42
	alfalfa	leaves and flowers	infusion or tincture internally	46
	aloe vera	gel	ointment	48
	angelica	roots	compress	50
	bay tree	leaves	compress, infused oil	56
	bitter aloe	whole leaf	ointment	60
	borage	leaves and flowers	compress, infused oil	64
	burdock	roots	tincture internally	68
	calendula	petals	compress, infused and essential oil	70
	catnip	leaves	poultice	77

MUSCLES, BONES AND JOINTS

Uses	Herb	Parts used	Method	Page
Arthritis and rheumatism cont.	chamomile	flowers	essential or infused oil in massage oil	78
	chickweed	leaves and flowers	decoction internally and as a compress	80
	chicory	leaves and flowers	infusion internally	81
	chilli	fruit	compress, infused oil	82
	comfrey	leaves, flowers and roots	infusion (leaves) in bath, infused oil (flowers, leaves and roots) in massage oil, tincture (roots) as massage	86
	coriander	seeds and leaves	infused oil (leaves and seeds) externally and compress (seeds)	88
	evening primrose	seeds and leaves	oil (seeds) internally, poultice (leaves) externally	100
	feverfew	leaves and flowers	tincture internally	104
	flax	seeds	oil internally	106
	ginger	rhizome	infusion or tincture internally	108
	gotu kola	leaves and stems	infusion and tincture internally, ointment externally	111
	horseradish	roots	poultice	116
	lemon balm	leaves and flowers	compress	121
	lemon grass	stems and leaves	ointment, infused oil as massage	123
	mustard	seeds	poultice	136
	nettle	leaves and flowers	infusion or tincture internally, compress	140
	oregano	leaves	infusion internally, poultice	128
	parsley	leaves	poultice	142
	peppermint	leaves and flowers	compress, or oil infusion or essential oil as massage	132
	rue	leaves and stems	ointment, poultice or infused oil externally	150
	soapwort	leaves and roots	ointment	157
	St John's wort	leaves and flowers	infused oil as massage	160
	tansy	leaves	compress	163
	thyme	leaves and flowers	essential oil added to massage oil	168
	yarrow	flowers	infusion internally	174
Backache	chilli	fruit	infused oil as massage	82
	horseradish	roots	poultice	116
	lemon grass	stems and leaves	ointment, infused oil as massage	123
	mustard	seeds	poultice	136
	rue	leaves and stems	ointment, poultice or infused oil externally	150

MUSCLES, BONES AND JOINTS

Uses	Herb	Parts used	Method	Page
Fractures	comfrey	leaves, flowers and roots	poultice	86
Gout	dandelion	leaves, flowers and stems	infusion internally (leaves, flowers, stems), juice (leaves)	92
	nettle	leaves and flowers	infusion internally, compress	140
	rue	leaves and stems	ointment	150
Osteo-arthritis (see Arthritis, page 272)	comfrey	flowers, leaves and roots	cream	86
	nettle	leaves and flowers	infusion internally	140
Osteoporosis	nettle	leaves and flowers	infusion internally	140
	parsley	roots	decoction internally to strengthen bones	142
	watercress	leaves	infusion internally	171
Sciatica	nettle	leaves and flowers	infusion internally, compress applied to affected areas	140
	rue	leaves and stems	ointment, poultice or infused oil externally	150
	St John's wort	flowers	cream	160
Sore and damaged muscles	calendula	petals	infused oil for sore muscles	70
	comfrey	leaves and flowers	cream for damaged muscles	86
	horseradish	roots	poultice for sore muscles	116
	lavender	flowers	infused or essential oil in massage oil	119
	lemon grass	stems and leaves	ointment, infused oil as massage for sore muscles	123
	oregano	leaves and flowers	poultice	128
	St John's wort	flowers	cream or infused oil for muscle cramp	160
Strains and sprains	borage	leaves and flowers	compress	64
	comfrey	leaves and flowers	poultice, infused oil	86
	lemon grass	stems and leaves	ointment, infused oil as massage	123
	nettle	leaves and flowers	compress applied to affected areas	140
	St John's wort	flowers	cream	160
Tendonitis	nettle	leaves and flowers	infusion internally, compress applied to affected areas	140

HEART, BLOOD, LYMPH SYSTEM AND CIRCULATION

Uses	Herb	Parts used	Method	Page
Anaemia	aloe vera	gel	infusion internally	48
	nasturtium	leaves	infusion internally	138
	nettle	leaves	juice	140
	watercress	leaves	infusion internally	171
Blood cleanser	burdock	roots	decoction or tincture internally	68
	clover	leaves and flowers	infusion internally	84
	lovage	leaves and seeds	infusion internally	126
	sorrel	leaves	infusion internally	158
	watercress	leaves	infusion internally	171
Blood vessels and bleeding	heartsease	flowers, leaves and roots	tincture internally to strengthen	114
	salad burnet	roots	poultice externally, decoction internally to stop bleeding	154
	selfheal	leaves and flowers	tincture or infusion internally or fresh leaves as poultice to stop bleeding	156
	yarrow	leaves	poultice to stop bleeding	174
Bruising	African wormwood	leaves	compress	44
	borage	leaves and flowers	compress	64
	bugle	whole plant	ointment, infused oil	66
	calendula	petals	ointment	70
	comfrey	flowers, leaves and roots	infused oil	86
	evening primrose	leaves	poultice	100
Diabetes and blood sugar	African potato	corms	infusion and decoction	42
	alfalfa	leaves and flowers	infusion and tincture internally	46
	cancer bush	leaves	infusion internally	73
	nettle	leaves and flowers	infusion internally	140
	stevia	leaves	infusion internally	162
Fluid retention	angelica	leaves	infusion internally	50
	cape gooseberry	fruit	infusion internally	74
	dandelion	leaves, flowers and stems	infusion (leaves, flowers, stem), juice (leaves)	92

HEART, BLOOD, LYMPH SYSTEM AND CIRCULATION

Uses	Herb	Parts used	Method	Page
Fluid retention cont.	horseradish	roots	infusion internally	116
	parsley	leaves	infusion or tincture internally	142
	sorrel	leaves	infusion internally	158
Heart ailments	dandelion	leaves	tincture internally	92
	evening primrose	seeds	oil internally to strengthen heart	100
	heartsease	flowers, leaves and roots	infusion internally	114
	motherwort	leaves and flowers	infusion internally	134
High blood pressure	African potato	corms	infusion and decoction internally	42
	heartsease	flowers, leaves and roots	infusion internally	114
	lemon balm	leaves and flowers	infusion or tincture internally	121
	motherwort	leaves and flowers	infusion internally	134
	rue	leaves and stems	infusion internally	150
	selfheal	flowers	decoction internally	156
	sorrel	leaves	infusion internally	158
	stevia	leaves	infusion internally	162
High cholesterol	African potato	corms	infusion and decoction internally	42
	alfalfa	leaves and flowers	infusion and tincture internally	46
	coriander	leaves and seeds	infusion internally	88
	dandelion	roots	decoction internally	92
	flax	seeds	oil internally	106
Poor circulation	angelica	leaves	infusion internally, cream externally	50
	chilli	fruit	infusion or tincture internally	82
	coriander	leaves and seeds	infusion internally	88
	evening primrose	seeds	oil internally	100
	ginger	rhizome	decoction or tincture internally	108
	gotu kola	leaves and stems	infusion or tincture internally	111
	heartsease	flowers, leaves and roots	infusion internally	114
	horseradish	roots	infusion internally	116
	mustard	seeds	infusion internally, foot bath	136
	nettle	leaves and flowers	infusion internally	140
	rosemary	leaves	infusion internally, cream externally	148
	rue	leaves and stems	infusion internally	150
	thyme	leaves and flowers	infusion internally	168

URINARY SYSTEM

Uses	Herb	Parts used	Method	Page
Kidney ailments	alfalfa	leaves and flowers	infusion and tincture	46
	cape gooseberry	fruit	infusion internally	74
	goldenrod	flowers and leaves	infusion or tincture internallly	110
Prostate	alfalfa	leaves and flowers	infusion and tincture internally	46
	evening primrose	seeds	oil internally	100
Urinary ailments	African potato	corms	infusion and decoction internally	42
	alfalfa	leaves and flowers	infusion and tincture internally	46
	blackberry	leaves	infusion internally for cystitis	62
	cancer bush	leaves	infusion internally	73
	chickweed	leaves and flowers	infusion internally	80
	dandelion	leaves, flowers and stems	infusion internally (leaves, flowers, stem), juice (leaves)	92
	fennel	roots	decoction internally	102
	goldenrod	flowers and leaves	infusion or tincture internallly	110
	savory	leaves	infusion or tincture internally for cystitis	155
	yarrow	flowers and leaves	tincture internally	174

MIND AND EMOTIONS

Uses	Herb	Parts used	Method	Page
Anxiety, stress and tension, anger and irritability	anise hyssop	leaves and flowers	infusion internally	52
	basil	leaves and flowers	infusion, steam inhalation, tincture, essential or infused oil in bath oil	54
	borage	leaves and flowers	infusion or juice internally	64
	calendula	petals and leaves	infusion internally, essential or infused oil in bath	70
	California poppy	whole plant	infusion internally	72
	cancer bush	leaves	infusion internally	73
	catnip	leaves	infusion internally	77
	chamomile	flowers	infusion internally, essential or infused oil in bath	78
	coriander	seeds	infusion internally, essential oil in burner	88
	dill	leaves and seeds	infusion or tincture internally	94
	evening primrose	seeds	oil internally	100
	gotu kola	leaves and stems	infusion internally	111

MIND AND EMOTIONS

Uses	Herb	Parts used	Method	Page
Anxiety, stress and tension, anger and irritability cont.	lavender	flowers	infusion or tincture internally, essential or infused oil as massage or in burner or bath oil	119
	lemon balm	leaves and flowering tips	infusion internally	121
	lemon grass	stems and leaves	infusion or tincture internally	123
	marjoram	leaves and flowers	infusion internally, essential or infused oil in an oil burner or bath oil	128
	motherwort	leaves and flowers	infusion internally	134
	mugwort	leaves and roots	infusion (leaves) or decoction (roots) internally	135
	oregano	leaves and flowers	essential or infused oil in oil burner or bath oil	128
	pelargonium	*P. graveolens* leaves	infusion internally, infused or essential oil as massage	144
	rue	leaves and stems	infusion internally	150
	selfheal	flowers	decoction internally	156
	St John's wort	leaves and flowers	infusion, tincture or extract internally	160
Depression	anise hyssop	leaves and flowers	infusion internally	52
	basil	leaves and flowers	infusion, steam inhalation, tincture, essential or infused oil in bath oil	54
	bergamot	leaves and flowers	infusion internally	58
	borage	leaves and flowers	infusion, syrup (flowers) or juice internally	64
	cancer bush	leaves	infusion internally	73
	chilli	fruit	infusion internally	82
	evening primrose	seeds	oil internally	100
	lavender	flowers	infusion or tincture internallly	119
	lemon balm	leaves and flowers	infusion or tincture internally, infused oil as massge	121
	lemon grass	stems and leaves	infusion or tincture internally	123
	motherwort	leaves and flowers	infusion internally, especially post childbirth	134
	pelargonium	*P. graveolens* leaves	infused or essential oil as massage	144
	rosemary	leaves	infusion internally	148
	sage	leaves	infusion internally	152
	salad burnet	leaves	infusion internally	154
	St John's wort	leaves and flowers	infusion, tincture or extract internally	160

MIND AND EMOTIONS

Uses	Herb	Parts used	Method	Page
Fatigue	basil	leaves and flowers	infusion, steam inhalation, essential or infused oil in bath oil	54
	chickweed	leaves and flowers	infusion internally	80
	gotu kola	leaves and stems	infusion internally	111
	sage	leaves	infusion internally	152
Forgetfulness and confusion	basil	leaves and flowers	infusion internally, steam inhalation	54
	gotu kola	leaves and stems	infusion and tincture internally	111
	rosemary	leaves and flowers	infusion internally	148
Grief	borage	leaves and flowers	infusion or juice internally	64
Hyperactivity	chamomile	flowers	infusion internally, essential or infused oil in bath	78
	gotu kola	leaves and stems	infusion internally	111
	lavender	flowers	infusion or tincture internallly	119
	lemon balm	leaves and flowers	infusion or tincture internally	121
	selfheal	flowers	decoction internally	156
Insomnia	California poppy	whole plant	infusion internally	72
	catnip	leaves	infusion internally	77
	chamomile	flowers	infusion internally	78
	dill	leaves and seeds	infusion or tincture internally	94
	lavender	flowers	infusion internally	119
	lemon balm	leaves and flowering tips	infusion internally	121
	motherwort	leaves and flowers	infusion internally	134
	mugwort	leaves and roots	infusion (leaves) or decoction (roots) internally	135
	pelargonium	*P. graveolens* leaves	infusion internally	144
	peppermint	leaves and flowers	infusion or tincture internally	132
	rose pelargonium	leaves and flowers	infusion internally, essential oil	144
	tarragon	leaves	infusion internally	164
Shock	chilli	fruit	infusion internally	82

EYES, EARS, THROAT, NOSE, MOUTH AND LUNGS

Uses	Herb	Parts used	Method	Page
Asthma	African potato	corms	infusion and decoction internally	42
	angelica	leaves	tincture internally	50
	cancer bush	leaves	infusion internally	73
	cape gooseberry	fruit	infusion internally	74
	chamomile	flowers	steam inhalation	78
	evening primrose	leaves and flowers	infusion internally	100
	horseradish	root	infusion internally	116
	lavender	flowers	essential or infused oil as a chest rub or steam inhalation	119
	lemon balm	leaves and flowers	infused oil as chest rub	121
	lovage	leaves and seeds	infusion internally	126
	marjoram	leaves and flowers	infusion internally	128
	mustard	seeds	infusion internally	136
	thyme	leaves and flowers	infusion internally	168
	wild garlic	leaves and flowers	infusion internally	173
	yarrow	flowers	infusion internally, steam inhalation	174
Breath	coriander	seeds	decoction to sweeten breath	88
	dill	seeds	chew dried seeds to sweeten breath	94
	lavender	flowers	infusion as a mouthwash	119
	oregano	leaves and flowers	infusion as a mouthwash	128
	peppermint	leaves and flowers	infusion or tincture as a mouthwash	132
	rosemary	leaves	infusion as a mouthwash	148
	tarragon	leaves	chew fresh leaves to sweeten breath	164
Bronchitis	African wormwood	leaves	infusion and tincture internally	44
	angelica	leaves and roots	tincture internally	50
	bergamot	leaves and flowers	infusion internally	58
	borage	leaves and flowers	infused oil as a chest rub	64
	cancer bush	leaves	infusion internally	73
	chamomile	flowers	steam inhalation	78
	elder	flowers	tincture internally	98
	houseleek	leaves	infusion internally	118
	lemon balm	leaves and flowers	infused oil as a chest rub	121
	nasturtium	leaves	infusion internally	138

QUICK GUIDE

EYES, EARS, THROAT, NOSE, MOUTH AND LUNGS

Uses	Herb	Parts used	Method	Page
Bronchitis cont.	oregano	leaves and flowers	infusion internally	128
	pelargonium	*P. graveolens* leaves	steam inhalation	144
	savory	leaves	infusion or tincture internally	155
	tea tree	leaves and branches	steam inhalation with infusion or essential oil	166
	thyme	leaves and flowers	infusion or syrup internally, infused or essential oil as a chest rub	168
	watercress	leaves	infusion internally	171
	wild garlic	leaves and flowers	infusion internally	173
	yarrow	flowers	infusion internally, steam inhalation	174
Colds and flu	African wormwood	leaves	infusion, tincture internally	44
	anise hyssop	leaves and flowers	infusion internally	52
	basil	leaves and flowers	steam inhalation, tincture internally	54
	cancer bush	leaves	infusion internally	73
	catnip	leaves and flowers	infusion internally	77
	chilli	fruit	infusion internally	82
	chives	leaves and flowers	infusion internally, to prevent and treat colds	83
	echinacea	flowers, leaves and roots	decoction internally	96
	elder	flowers	infusion or tincture internally	98
	fennel	leaves and seeds	essential and infused oil as a chest rub	102
	ginger	rhizome	decoction or tincture internally	108
	goldenrod	leaves and flowers	infusion or tincture internally	110
	ground ivy	leaves and flowers	infusion internally	113
	horseradish	roots	infusion internally	116
	lemon balm	leaves and flowers	infusion or tincture internally	121
	marjoram and oregano	leaves and flowers	infusion internally	128
	mustard	seeds	infusion internally, poultice	136
	nasturtium	leaves	infusion internally	138
	savory	leaves	infusion or tincture internally	155
	tea tree	leaves and branches	steam inhalation with infusion or essential oil	166
	wild garlic	leaves and flowers	infusion internally	173

EYES, EARS, THROAT, NOSE, MOUTH AND LUNGS

Uses	Herb	Parts used	Method	Page
Expectorant	angelica	leaves	tincture internally	50
	basil	leaves and flowers	syrup	54
	borage	flowers	syrup	64
	thyme	leaves and flowers	tincture or syrup internally	168
	watercress	leaves	infusion internally	171
Eyes	aloe vera	gel	apply to outside of eyelid for conjunctivitis	48
	bitter aloe	whole leaf	infusion as wash for conjunctivitis	60
	calendula	petals	compress for sore eyes	70
	chamomile	flowers	infusion as wash for strained eyes or conjunctivitis	78
	chicory	leaves and flowers	infusion internally	81
	cornflower	flowers	infusion as an eyewash	90
	elder	flowers	infusion as an eyewash	98
	fennel	seeds	infusion as an eyewash	102
	ground ivy	leaves and flowers	weak infusion as an eyewash	113
	parsley	leaves	compress for sore eyes	142
	raspberry	leaves	wash for tired eyes	62
	rue	leaves and stems	infusion as an eyewash, internally to improve eyesight	150
	selfheal	leaves and flowers	weak infusion as a wash for hot tired or infected eyes	156
Mouth ulcers and infections	bergamot	leaves and flowers	infusion as a mouthwash	58
	blackberry and raspberry	leaves and flowers	infusion as a mouthwash	62
	calendula	petals and leaves	infusion as a mouthwash	70
	chamomile	flowers	infusion as a mouthwash	78
	coriander	seeds	infusion as a mouthwash	88
	echinacea	flowers, leaves and root	decoction as a mouthwash	96
	elder	flowers	infusion as a mouthwash	98
	fennel	seeds	infusion as a mouthwash	102
	ground ivy	leaves and flowers	infusion as a mouthwash	113
	houseleek	leaves	infusion as a mouthwash	118
	marjoram and oregano	leaves and flowers	infusion as a mouthwash	128
	rosemary	leaves	infusion as a mouthwash	148
	sage	leaves	infusion as a mouthwash	152

EYES, EARS, THROAT, NOSE, MOUTH AND LUNGS

Uses	Herb	Parts used	Method	Page
Mouth ulcers and infections cont.	selfheal	leaves and flowers	infusion or tincture as a mouthwash	156
	tea tree	leaves and branches	infusion or essential oil mixed with apple cider vinegar as a mouthwash	166
	thyme	leaves and flowers	infusion or tincture as a mouthwash	168
	watercress	leaves	poultice or tincture as a mouthwash	171
Nosebleeds	salad burnet	roots	decoction internally	154
	selfheal	leaves	poultice in nostril or infusion or tincture internally	156
	yarrow	leaves	poultice in nostril	174
Sinusitis	basil	leaves and flowers	steam inhalation	54
	elder	flowers	infusion or tincture internally	98
	goldenrod	flowers and leaves	infusion internally	110
	ground ivy	leaves and flowers	steam inhalation	113
	horseradish	root	infusion internally	116
	marjoram	leaves and flowers	infusion internally	128
	pelargonium	*P. betulinum* leaves	steam inhalation	144
	peppermint	leaves and flowers	steam inhalation with fresh leaves or essential oil	132
	tea tree	leaves and branches	steam inhalation with infusion or essential oil	166
Sore throat and cough	angelica	leaves	tincture internally for cough	50
	anise hyssop	leaves and flowers	infusion internally and syrup for cough	52
	basil	leaves and flowers	syrup for cough	54
	bergamot	leaves and flowers	infusion internally for sore throat or cough	58
	blackberry	leaves	infusion as a gargle	62
	borage	leaves and flowers	infusion internally, syrup (flowers) for cough	64
	calendula	petals and leaves	infusion as a gargle, tincture internally	70
	chilli	fruit	tincture as a gargle	82
	clover	flowers and leaves	syrup for sore throat and cough	84
	echinacea	flowers, leaves and roots	decoction as a gargle	96
	elder	flowers, berries	infusion as a gargle (flowers), decoction and syrup as cough mixture (berries)	98
	evening primrose	leaves and flowers	infusion internally for cough	100
	fennel	seeds	infusion as a gargle	102
	flax	seeds	infusion internally for sore, swollen throat	106
	ginger	rhizome	decoction or tincture internally for cough or sore throat	108

EYES, EARS, THROAT, NOSE, MOUTH AND LUNGS

Uses	Herb	Parts used	Method	Page
Sore throat and cough cont.	goldenrod	flowers and leaves	infusion or tincture internally or as a gargle	110
	ground ivy	leaves and flowers	infusion as a gargle	113
	heartsease	flowers, leaves and root	syrup or tincture internally for cough	114
	houseleek	leaves	infusion as a gargle	118
	lovage	leaves, seeds and roots	infusion (leaves and seeds) or decoction (root) internally for sore throat	126
	marjoram and oregano	leaves and flowers	infusion internally or gargle for cough	128
	mustard	seeds	infusion internally for cough	136
	nasturtium	leaves	infusion internally for cough and sore throat	138
	pelargonium	*P. graveolens* leaves	infusion as a gargle for sore throat	144
	raspberry	leaves, berries	infusion as gargle (leaves), vinegar in cough mixture (berries)	62
	rocket	leaves and seeds	syrup for coughs	146
	sage	leaves	infusion internally or gargle for cough	152
	savory	leaves	infusion or tincture internally for sore throat	155
	selfheal	leaves and flowers	infusion or tincture as a gargle	156
	tea tree	leaves and branches	infusion or essential oil mixed with apple cider vinegar as a gargle for sore throat, steam inhalation with infusion or essential oil for cough	166
	thyme	leaves and flowers	infusion or tincture as a gargle for sore throat	168
	watercress	leaves	infusion internally for cough	171
	wild garlic	leaves and flowers	infusion internally for cough	173
Tonsillitis	calendula	petals and leaves	tincture internally	70
Toothache and teeth	California poppy	fresh root	poultice on sore teeth	72
	chamomile	flowers	infusion as a mouthwash	78
	marjoram and oregano	leaves and flowers	essential and infused oil on affected tooth	128
	nettle	leaves and flowers	infusion to strengthen teeth	140
	stevia	leaves	infusion as a mouthwash to reduce tooth decay	162
	tarragon	leaves	chew fresh leaves	164
	watercress	leaves	infusion internally to strengthen teeth, as a mouthwash for gum infections	171

IMMUNE SYSTEM, TONICS AND ALLERGIES

Uses	Herb	Parts used	Method	Page
Allergies	African potato	corms	infusion and decoction internally	42
	cape gooseberry	fruit	infusion internally	74
	chamomile	flowers	steam inhalation	78
	echinacea	flowers, leaves and root	decoction internally	96
	elder	flowers	infusion internally	98
	goldenrod	leaves and flowers	infusion or tincture internally	110
	lovage	leaves and seeds	infusion internally	126
	marjoram and oregano	leaves and flowers	infusion internally	128
	nettle	leaves and flowers	infusion internally	140
	sage	leaves	infusion internally	152
	thyme	leaves and flowers	infusion internally	168
	yarrow	flowers	infusion internally	174
Antiviral	African potato	corms	infusion and decoction internally	42
	aloe vera	gel	infusion and decoction internally	48
	California poppy	whole plant	infusion internally	72
	cancer bush	leaves	infusion or tincture internally	73
	cape gooseberry	fruit and leaves	infusion internally	74
	clover	flowers and leaves	infusion internally	84
	echinacea	leaves, flowers and roots	infusion or decoction internally	96
	elder	flowers	infusion internally	98
	lemon balm	leaves and flowers	infusion internally	121
	marjoram and oregano	leaves and flowers	infusion internally	128
	nettle	leaves and flowers	infusion internally	140
General tonic	alfalfa	leaves, flowers, young shoots and sprouts	juice or whole plant	46
	aloe vera	gel	infusion and decoction internally	48
	borage	leaves and flowers	infusion internally	64
	cancer bush	leaves	infusion internally	73
	chicory	leaves and flowers	infusion internally	81
	cornflower	leaves and flowers	infusion internally	90
	nettle	leaves and flowers	juice	140
	parsley	leaves	infusion internally	142
	rocket	leaves and seeds	infusion internally	146

IMMUNE SYSTEM, TONICS AND ALLERGIES

Uses	Herb	Parts used	Method	Page
General tonic cont.	sorrel	leaves	infusion internally	158
	tarragon	leaves	infusion internally	164
	watercress	leaves	infusion internally	171
Immune system strengthening	African potato	corms	infusion and decoction internally	42
	angelica	leaves	infusion and decoction internally	50
	cancer bush	leaves	infusion internally	73
	echinacea	leaves, flowers and roots	decoction internally	96
	heartsease	leaves, flowers and roots	infusion internally	114
	nettle	leaves and flowers	infusion internally	140
	thyme	leaves and flowers	infusion internally	168

GYNAECOLOGICAL

Uses	Herb	Parts used	Method	Page
Menopause	alfalfa	leaves and flowers	infusion and tincture internally	46
	angelica	roots	decoction internally	50
	calendula	petals and leaves	infusion internally	70
	clover	flowers and leaves	infusion internally	84
	evening primrose	seeds	oil internally for menopause symptoms	100
	flax	seeds	ground seeds	106
	parsley	leaves	infusion or tincture internally to treat emotional imbalance	142
	sage	leaves	tincture internally for night sweats and hot flushes	152
	St John's wort	leaves and flowers	infusion, tincture and extract internally for mood swings	160
	yarrow	flowers	infusion internally for irregular menstruation	174
Menstrual	alfalfa	leaves and flowers	infusion and tincture internally, for irregular menstruation	46
	angelica	roots	decoction internally	50
	calendula	petals and leaves	infusion internally, for cramps	70
	caraway	seeds	infusion internally, for cramps	76
	chamomile	flowers	infusion internally, for cramps	78

QUICK GUIDE

GYNAECOLOGICAL

Uses	Herb	Parts used	Method	Page
Menstrual cont.	evening primrose	seeds	oil internally, for premenstrual symptoms	100
	feverfew	leaves and flowers	fresh leaf (headache), infusion of leaves and flowers (period pains)	104
	lovage	leaves and seeds	infusion internally, to encourage menstruation	126
	motherwort	leaves and flowers	infusion internally, to regulate menstruation and ease cramps	134
	mugwort	leaves and roots	infusion (leaves) or decoction (roots) internally, to regulate menstruation (especially in young women)	135
	oregano	leaves and flowers	infusion internally, to promote and regulate menstruation	128
	parsley	leaves	infusion or tincture internally, to replace iron	142
	pelargonium	*P. graveolens* leaves	infused or essential oil as massage	144
	rue	leaves and stems	infusion internally, to reduce cramps and induce menstrual flow	150
	selfheal	leaves and flowers	infusion or tincture internally, to reduce heavy menstruation	156
	St John's wort	leaves and flowers	infusion, tincture and extract internally, for premenstrual symptoms	160
	yarrow	flowers and leaves	tincture internally, to regulate menstruation	174
Thrush	calendula	petals	infusion as a douche	70
	goldenrod	flowers and leaves	infusion as a douche	110
	tea tree	leaves and branches	infusion as a douche	166
	thyme	leaves and flowers	infusion as a douche	168

PREGNANCY AND CHILDBIRTH

Uses	Herb	Parts used	Method	Page
Breast-feeding	borage	leaves and flowers	infusion, to stimulate lactation	64
	calendula	petals	ointment, infused oil, essential oil or compress for sore nipples	70
	dill	seeds	infusion, to stimulate lactation	94
	fennel	seeds	infusion, to stimulate lactation	102

PREGNANCY AND CHILDBIRTH

Uses	Herb	Parts used	Method	Page
Breast-feeding cont.	milk thistle	leaves, ground seeds and flowers	infusion (leaves and seeds) and tincture (leaves, seeds and flowers) internally, to increase lactation	130
	nettle	leaves and flowers	infusion internally, to increase lactation	140
	parsley	leaves	infusion or tincture internally, to increase lactation and to replace iron	142
	raspberry	leaves	infusion internally, to increase lactation	62
Morning sickness	ginger	rhizome	decoction or tincture internally	108
Uterine tonic	motherwort	leaves and flowers	infusion internally	134
	raspberry	leaves	infusion internally in last 6–8 weeks of pregnancy	62

BRAIN AND NERVES

Uses	Herb	Parts used	Method	Page
Dizziness	feverfew	leaves and flowers	infusion internally	104
	selfheal	flowers	decoction internally	156
Headache	basil	leaves and flowers	steam inhalation	54
	bay tree	leaves and flowers	infused oil on temples	56
	coriander	whole plant	essential oil in a burner, compress (seeds)	88
	feverfew	leaves and flowers	eat fresh leaves, tincture internally	104
	lavender	flowers	infusion internally, essential oil and infused oil externally	119
	lemon balm	leaves and flowers	infusion internally	121
	marjoram and oregano	leaves and flowers	infusion internally	128
	peppermint	leaves and flowers	compress, infused oil or essential oil on temples	132
	rosemary	leaves and flowers	essential oil and infused oil on temples	148
Migraine	angelica	roots	compress	50
	bay tree	leaves and flowers	infused oil on temples	56
	feverfew	leaves and flowers	eat fresh leaves, tincture internally	104
	tansy	leaves	compress	163

BRAIN AND NERVES

Uses	Herb	Parts used	Method	Page
Neuralgia	California poppy	whole plant	infusion internally	72
	chamomile	flowers	infusion internally, ointment	78
	chilli	fruit	infused oil in a massage oil	82
	coriander	seeds	compress, infusion internally	88
	gotu kola	leaves and stems	infusion or tincture internally	111
	lavender	flowers	infusion internally	119
	peppermint	leaves and flowers	ointment, oil infusion or essential oil externally	132
	rue	leaves and stems	ointment, poultice or infused oil externally	150
	St John's wort	flowers	cream or infused oil externally, infusion internally	160
	tansy	leaves	compress	163
Shingles	aloe vera	leaves, gel inside leaves	fresh gel	48
	bitter aloe	leaves, gel inside leaves	fresh gel	60
	bulbine	leaves, gel inside leaves	fresh gel	67
	rue	leaves and stems	ointment, poultice or infused oil externally	150
	St John's wort	flowers	cream or infused oil	160

OTHER

Uses	Herb	Parts used	Method	Page
Antioxidant	basil	leaves	infusion internally	54
	blackberry and raspberry	leaves and fruit	infusion internally	62
	California poppy	leaves, flowers and seeds	infusion internally	72
	cancer bush	leaves	infusion internally	73
	cape gooseberry	leaves and fruit	infusion internally	74
	echinacea	flowers, leaves and root	decoction or infusion internally	96
	marjoram and oregano	leaves and flowers	infusion internally	128
	milk thistle	leaves, ground seeds and flowers	infusion internally	130
	rocket	leaves, flowers and seeds	infusion internally	146
	watercress	leaves	infusion internally	171

OTHER

Uses	Herb	Parts used	Method	Page
Cancer	cancer bush	leaves	infusion internally	73
	clover	flowers and leaves	tincture internally	84
	dandelion	roots	decoction, to help deal with chemotherapy	92
	milk thistle	leaves, ground seeds and flowers	tincture, to reduce side effects of chemotherapy	130
	watercress	leaves	infusion internally, to prevent lung and prostate cancers	171
Infections and fever	anise hyssop	leaves and flowers	infusion internally, to promote sweating	52
	burdock	seeds	decoction internally, to reduce fever	68
	calendula	petals and leaves	infusion, to treat infections	70
	cancer bush	leaves	infusion internally	73
	catnip	leaves	infusion internally, to reduce fever	77
	chickweed	roots	decoction internally, to reduce fever	80
	chilli	fruit	infusion internally, to promote sweating	82
	echinacea	flowers, leaves and root	decoction internally, to reduce fever	96
	elder	flowers	infusion internally, to reduce fever	98
	ginger	rhizome	decoction or tincture internally, to promote sweating	108
	ground ivy	leaves and flowers	infusion internally	113
	lemon grass	stems and leaves	infusion or decoction internally, to reduce fever	123
	mugwort	leaves and roots	infusion (leaves) or decoction (roots) internally	135
	peppermint	leaves and flowers	infusion or tincture internally, to promote sweating	132
	yarrow	flowers and leaves	infusion internally	174
Inflammation	cancer bush	leaves	infusion internally	73
	cape gooseberry	leaves	poultice	74
	echinacea	flowers, leaves and roots	decoction internally	96
Pain	California poppy	whole plant	infusion, to ease pain	72
	echinacea	flowers, leaves and roots	decoction, to ease pain	96

BIBLIOGRAPHY

Aldworth, Diane. *The Wonders of Weeds* (Direct 2U DTP, 2003)
Bremness, Lesley. *The Essential Herbs Handbook* (Duncan Baird Publishers, 2009)
Chiej, Roberto. *The Macdonald Encyclopaedia of Medicinal Plants* (Macdonald & Co, 1984)
Clevely, Andi. *Herbs. A User's Guide and Identifier* (Hermes House, 2000)
Culpeper, Nicholas. *Culpeper's Complete Herbal* (Chartwell Books, 1985)
DK Publishing. *Neal's Yard Remedies* (Dorling Kindersley, 2011)
Farmer-Knowles, Helen. *The Healing Plants Bible* (Godsfield, 2010)
Frayne, Margie. *Help Yourself to Health* (Margie Frayne, 2005)
French, Jackie. *Natural Solutions* (ACP Publishing, 1999)
Houdret, Jessica. *The Practical Illustrated Home Herbal Doctor* (Hermes House, 2009)
McVicar, Jekka. *Grow Herbs* (Dorling Kindersley, 2010)
McVicar, Jekka. *Jekka's Complete Herb Book* (Kyle Cathie, 2007)
Mnimh, Penelope Ody. *The Herb Society's Complete Medicinal Herbal* (Dorling Kindersley, 1995)
Newton, Anna. *Herbs for Home Treatment* (Green Books, 2009)
Reader's Digest Southern Africa. *Ouma's Home Remedies.* (Heritage Publishers, 2002)
Reader's Digest. *The Ultimate Book of Herbs* (The Reader's Digest Association, 2009)
Reichardt, Irmela. *Natural Gardening* (Delta Books, 1993)
Roberts, Margaret. *A – Z of Herbs* (Struik Publishers, 2000)
Shaw, Non. *Herbalism. An Illustrated Guide* (Element Books, 1998)
Shealy, Norman C. *The Illustrated Encyclopaedia of Healing Remedies* (Harper Collins, 2002)
The Royal Horticultural Society. *Encyclopaedia of Herbs and their Uses* (Dorling Kindersley, 2002)
Treben, Maria. *Health from God's Garden* (Healing Arts Press, 1986)
Treben, Maria. *Health through God's Pharmacy* (Ennsthaler Steyr, 2009)
Weiss, Gaea and Shandor. *Growing and Using Healing Herbs* (Wings Books, 1999)
Wong, James. *Grow Your Own Drugs. Easy Recipes for Natural Remedies and Beauty Treats* (Collins, 2009)
Wong, James. *Grow Your Own Drugs: A Year with James Wong* (Collins, 2010)

ACKNOWLEDGEMENTS

Zirkia Swart from Mountain Herb Estate for her generosity, knowledge and magical herb garden.
Jenny Slabber, Claire Slabber, Heidi Weeks and Diana Smith from Talborne Organics, for their support and passion for organic growing in South Africa.
Ceri Prenter, Ingeborg Pelser, Jeremy Boraine, Claire Richards and the rest of the team at Sunbird Publishers for their enthusiasm and energy.
Fresh Earth Food Store for their help and healthy goodies.
Linda de Luca of Random Harvest Indigenous Nursery for her generosity.
Hloniphani Valoyi for his energy, creativity and always being willing to try out my gardening ideas.
Mary-Jane Harris and her team at South African *Garden and Home* magazine for all their support.

GENERAL INDEX

Abscesses 193, **260**
Acacia trees 179, 183
Acne see Pimples
African potato (*Hypoxis hemerocallidea*) **42–3**
African wormwood (*Artemisia afra*) 29, **44–5**
  used in recipes 213, 215, 245, 254, 257, 259
Ageing 216, 219, 227, **260**
Alcohol 195
Alfalfa (*Medicago sativa*) 31, **46–7**
  used in recipes 201, 206, 208, 218, 223, 238, 243, 247, 254, 257
Allergies **285**
Aloe vera (*Aloe vera*) **48–9**, 185
  used in recipes 189, 222, 228, 251, 252, 256
Anaemia 247, **275**
Angelica (*Angelica archangelica*) 30, **50–1**
  used in recipes 235, 243, 248
Anger 246, **277–8**
Anise hyssop (*Agastache foeniculum*) 36, **52–3**
  used in recipes 207, 232, 237, 243, 245, 247
Annual herbs 21, 27, 34
Antioxidant properties 183, 185, **289**
  in remedies 233, 247
Antiviral properties **285**
  in remedies 200, 230–38
Ants see Pests
Anxiety 243, 246, 247, 256, **277–8**
Aphids see Pests
Appetite (stimulate and repress) **267**
Artemisia see African wormwood
Arthritis 238, 239, 240, **272–3**
Arugula see Rocket
Aspirin (acetysalicylic acid) 177
Asthma 231, **280**
Assassin bugs see Insects, beneficial
Avocado oil 184, 185

Backache 238, 239, **273**
Balms 182, 192, 193
  see also Herbal Recipes Index
Basil (*Ocimum basilicum*) 15, 25, 27, 31, 38, **54–5**
  used in the recipes 181, 192, 208, 210, 246, 251
Bay tree (*Laurus nobilis*) 29, 36, **56–7**
  used in recipes 234, 239, 240, 245
Beauty, herbs for 181, 216–29
Bee balm see Bergamot, Lemon balm
Bees see Insects, beneficial
Beeswax 182, 188, 189
Beneficial herbs 29, 30–1
Beneficial insects see Insects, beneficial
Bergamot (*Monarda* species) **58–9**
  used in recipes 212, 246
Birds 29, 60, 75, 80, 153
Bites 251, 256, **266–67**
Bitter aloe (*Aloe ferox*) **60–1**
Blackberry (*Rubus fruticosus*) **62–3**
Bleeding 248, **275**
Blood cleanser 247, **275**
Blood pressure, high 247, **276**
Blood sugar see Diabetes
Blood vessels **275**

Boils 193, **260**
Borage (*Borago officinales*) 30, **64–5**
  used in recipes 206, 231, 236, 246, 247, 252
Borax 182, 185, 188, 189
Bottles and jars 186
  sterilising 188
Breastfeeding 180, **287–78**
Breath 225, **280**
Bronchitis 231, 235, 237, **280–81**
Bruising 239, **275**
Bugle (*Ajuga reptans*) 30, **66**
Bulbine (*Bulbine fructescens*) **67**
  used in recipes 236, 256
Burdock (*Arctium lappa*) **68–9**
  used in recipes 241, 243
Burns (and sunburn) 183, 190, 251, **260, 265–66**
Butterflies see Insects, beneficial

Cabbage moth see Pests
Calcium-rich plants 31, 86, 92, 93, 141, 171
Calendula (*Calendula officinalis*) 30, **70–1**, 179
  used in recipes 201, 216, 218, 219, 220, 221, 228, 229, 236, 250, 251, 252, 256
California poppy (*Eschscholtzia californica*) **72**
  used in recipes 241, 245, 246
Cancer **290**
Cancer bush (*Sutherlandia frutescens*) **73**
Cape gooseberry (*Physallis peruviana*) **74–5**
Caraway (*Carum carvi*) **76**, 95
  used in recipes 242
Carpenter's herb see Selfheal
Caterpillar see Pests
Catnip (*Nepeta cataria*) **77**
  used in recipes 245, 256, 257
Cellulite 248, **261**
Chamomile (*Anthemis nobilis* and *Matricaria recutita*) 27, 30, **78–9**
  used in recipes 216, 227, 228, 231, 234, 239, 241, 242, 243, 245, 246, 251, 252, 256, 259
Chequerboard garden design 15
Chickweed (*Stellaria media*) **80**
  used in recipes 239, 257
Chicory (*Cichorium intybus*) **81**
Chilblains **261**
Chilli (*Capsicum frutescens*) **82**
  used in recipes 203, 205
Chives (*Allium schoenoprasum*) 31, **83**
  used in recipes 202, 207, 208
Chlorophyll-rich plants 46, 140
Cholesterol, high 247, **276**
Cilantro see Coriander; Vietnamese coriander
Cinnamon 185, 190
Circulation, poor 247, 248, **276**
Citric acid/vitamin C powder 185
Clover 30, **84–5**
  used in recipes 201, 243, 254, 256, 259
Cockroach see Pests
Cocoa butter 183
Coconut oil 183, 192
Colds and flu 281, 230–33, 235–38
Cold sores 250, **261**

Colic 242, 243, **271–72**
Comfrey (*Symphytum officinale*) 31, 36, **86–7**
  used in recipes 251, 252, 257, 258
Companions, herbal **29–31**
  for beans 70, 103, 139
  for broccoli 45, 70, 139
  for cabbage (and brassica family) 31, 45, 59, 70, 78, 89, 95, 104, 115, 123, 132, 139, 145, 153
  for carrots 95, 107, 115
  for cauliflower 45, 70, 104
  for cucumber 55, 95, 132, 139
  for cucurbits 55
  for eggplant 31, 89
  for lettuce 95, 99, 104, 115, 147
  for onions 78, 95, 115, 147
  for peas 70, 76
  for potatoes 31, 86, 95, 107
  for pumpkin and squash 55, 121, 139
  for radish 147
  for strawberries 45, 65
  for tomatoes 29, 31, 55, 59, 65, 86, 89, 103, 121, 132, 139
Compost 18, 22, 27, 32
  vermicompost (earthworm castings) 22, 32
Compresses 193, 227
Coneflower see Echinacea
Confusion 246, **279**
Constipation **267–68**
Container herb gardens 15, 32, 34, 43
Coriander (*Coriandrum sativum*) 15, 21, 27, 30, 38, **88–9**
  see also Vietnamese coriander
  used in recipes 201, 202, 203, 205, 206, 207, 208, 213, 239, 241, 242
Cornflower (*Centaurea cyanus*) **90–1**
  used in recipes 227
Cottage-style herb garden 15, 50
Coughs 190, 230, 231, 232, 233, 235, 237, **283–84**
Cramps see Stomach cramps; Muscles, sore
Cream
  aqueous 182
  basic body cream 182, 188, 189
  lotions 188
  see also Herbal Recipe index
Cuttings, to propagate herbs 21, 23, 25
Cuts 251, 252, 256, **266–67**
Cutworm see Pests
Cystitis see Urinary ailments

Dandelion (*Taraxacum officinale*) 34, **92–3**, 177
  used in recipes 201, 206, 243, 257
Decoctions 189, 193
Depression 241, 247, **278**
Design of garden 12–15, 21
Detergents 182, 184
  natural cleaners 212–14
Detoxifying 206, 243, **268**
Diabetes and blood sugar 247, **275**
Diagnostics 180–81
Diarrhoea **268**
Digestion 242, 243, **269–70**
Dill (*Anethum graveolens*) 21, 29, 30, **94–5**, 242,
Diseases of herbs 17, 27, 31
Division, to propagate herbs 21, 23, 25, 26

Dizziness **288**
Drying herbs 36, 189

Earthworms see Worms
Echinacea (*Echinacea purpurea*) 30, **96–7**
    used in recipes 177, 190, 230, 232, 233, 234, 237
Eczema 182, 183, 184, 251, 252, **261–62**
Elder (*Sambucus nigra*) **98–9**, 254
    used in recipes 190, 200, 203, 204, 230, 232, 233, 234, 235, 237, 251, 254, 259
Evening primrose (*Oenothera biennis*) **100–1**
    used in recipes 184, 216, 220, 227, 249
Expectorant 230, 235, **282**
Eyes 216, 227, 228, **282**

Fatigue 246, **279**
Feeding herbs 27, 258
Fennel (*Foeniculum vulgare*) 21, 30, 76, 89, 95, **102–3**
    used in recipes 201, 206, 231, 242, 243, 254, 257
Fenugreek 30
Fertiliser 22, 27, 86, 87, 258
    see also Organic fertiliser
Fever **290**
Feverfew (*Tanacetum parthenium*) 29, 30, **104–5**
    used in recipes 213, 215, 257, 259
Flatulence (wind) 242, 243, **270**
Flax (*Linum usitatissimum*) **106–7**
    seed oil 107, 192
    used in recipes 192, 243, 249, 255
Flies see Pests
Florence fennel 102
Flu see Colds and flu
Fluid retention **275–76**
Forgetfulness 246, **279**
Fractures **274**
Fungal conditions 221, 224, 259, **262**
Fungi 17, 185, 188, 192

Gall bladder **270**
Gamma-linolenic acid (GLA) 65, 101
Garden
    designs **12–15**
    no-dig 19
    site for a 15–17
    soil for 17–18
    themes 21
Garlic 30, 190
    and sunflower oil spray 31
Gastro-enteritis **268**
Gelatine 183
Geranium essential oil 185
Germination rate 21, 22
Ginger (*Zingiber officinale*) **108–9**
    used in recipes 55, 190, 203, 205, 206, 230, 231, 232, 234, 235, 237, 238, 240, 242, 244, 247, 248
Glycerine 183, 189
Goldenrod (*Solidago virgaurea*) 30, **110**
    used in recipes 245
Goldenseal root 185
Gotu kola (*Centella asiatica*) **111–12**
    used in recipes 201, 206, 216, 218, 220, 223, 239, 241, 246, 248, 252
Gout 238, 239, 240, **274**
Green tea extract and powder 185

Grief **279**
Ground ivy (*Glechoma hederacea*) 113
Gum Arabic 183

Haemorrhoids 184, **271**
Hair 218, 223, 263
Harmful insects see Insects, harmful; Pests
Harvesting herbs 34
Headaches 193, 196, 239, **288**
Head lice 143
Healing herbs, quick guide **260–90**
Health, herbs for 177–78
    see also Herbal Recipes Index
Health from God's Garden (Maria Treben) 179
Heart ailments 247, **276**
Heartsease (*Viola tricolor*) **114–15**
    used in recipes 237, 247
Hen and chicks see Houseleek
Herbal companions see Companions
Herbal pharmacy **177–81**
    stocking of **182–86**
Herbal preparations **188–96**
Herbal sprays for the garden see Sprays
Herb wheel design 13
Home, herbs for 181, **212–15**
Honey 25, 183, 185
    herbal 189–90, 207, 252
    herbal infused 38
    herb honey balls 195
    herbs and spices for 190
    preserving in 38
    syrup 194–95
    to encourage root growth 25
Horseradish (*Armoracia rusticana*) 31, **116–17**
    used in recipes 201, 240, 245, 259
Houseleek (*Sempervivum tectorum*) 118
Hoverfly see Insects, beneficial
Humus 17, 18, 19
Hyperactivity **279**

Immune system strengthening 230, 235, 237, **286**
Infections 238, 245, 259, **290**
Inflammation 227, 228, 234, 238, 241, 242, 245, 252, 254, **290**
Infusions 190, 192, 193, 194, 195
    cold oil 193
    hot oil 192
Ingredients, basic **182–84**
Insecticides 29, 259
    natural see Sprays
Insects, beneficial 29
    assassin bugs 30
    bees 29, 30, 58, 60, 66, 85, 87, 96, 104
    butterflies 29, 30, 66, 96, 104, 113, 140, 156, 173
    herbs to attract 29, 30, 43, 47, 51, 58, 70, 76, 89, 91, 95, 103, 110, 119, 127, 134, 142, 145, 149
    hoverflies 95, 104, 154
    lacewings 110
    ladybirds 30, 31, 95, 103, 163
    spiders 29, 30
    terminators (bug-eating insects) 30
    wasps 29, 30, 76
Insects, harmful
    herbs to repel 29, 45, 55, 57, 59, 65, 70, 83, 99, 104,

119, 122, 123, 124, 133, 139, 145, 149, 151, 153, 163, 169, 173, 175
    insect-repelling recipes 157, 213, 215, 254, 255, 259
    trap herbs 29
    see also Pests
Insomnia 241, 245, **279**
Irritability 246, 249, **277–78**
Irritable bowel syndrome 242, 243, **271**
Italian garden theme 21

Jojoba oil 184, 185, 192
Juicing 190, 192

Karité tree (*Vitellaria paradoxa*) 184
Kidney ailments 206, 243, **277**
Kitchen, herbs in 181
Knitbone see Comfrey
Knot garden design 13

Labyrinth garden design 13, 14, 15
Lace wings see Insects, beneficial
Ladybirds see Insects, beneficial
Lavender (*Lavandula angustifolia*) 13, 15, 21, 23, 27, 29, 30, 36, 55, **119–20**
    used in recipes 190, 213, 214, 216, 219, 220, 221, 224, 227, 229, 231, 239, 241, 245, 246, 250, 251, 252, 255, 257, 259
Lawn chamomile see Chamomile
Layering, to propagate herbs 21, 23, 25
Legume family 30–1, 47, 85
    symbiotic relationship with bacteria 30–1
Lemon balm (*Melissa officinalis*) 30, **121–22**
used in recipes 190, 207, 215, 219, 226, 231, 232, 233, 247, 250, 255
Lemon grass (*Cymbopogon citratus*) **123–24**
    essential oil, as preservative 185
    used in recipes 190, 208, 214, 215, 219, 226, 255, 259
Lemon verbena (*Aloysia triphylla*) 23, 29, 36, **125**
    used in recipes 190, 208, 213, 215, 219, 223, 226
Linseed oil (from flax seed) 106
Liver problems 206, 243, **271**
Liquorice 190
Lovage (*Levisticum officinale*) **126–27**
    used in recipes 201

Maintenance, of herbs 27, 34
Manure 18, 19, 31, 32
Marjoram (*Marjorana hortensis*) 30, 31, 36, **128–29**
    used in recipes 190, 210
Medicinal garden theme 21
Medicinal herbs 180–81
Medieval garden theme 21
Mediterranean herbs 56, 119, 129, 147, 149, 150, 153, 168
    fungal problems 31
    garden theme 21
    oils 15
Melissa see Lemon balm
Menopause 249, **286**
Menstrual ailments 249, **286–87**
Mexican herbs garden theme 21
Migraine 239, **288**
Mildew 55, 259

Milk thistle (*Silybum marianum*) **130–31**, 206, 243
Mint (*Mentha* species) 29, 30, 31, 36, **132–33**
   see *also* Peppermint
   used in recipes 202, 204, 207, 208, 211, 214, 220, 239, 242, 243, 251
Morning sickness 244, **288**
Mosquitoes see Pests
Motherwort (*Leonurus cardiaca*) **134**
   used in recipes 245, 247
Mould 159, 185, 190, 192, 213
Mouth ulcers and infections **282–83**
Mugwort (*Artemisia vulgaris*) **135**, 145
Mulch 18, 27, 29, 31, 56, 86
Muscles, sore and damaged 239, **274**
Mustard (*Brassica* var. *juncea, alba, nigra*) 30, **136–37**
   used in recipes 239, 240, 242, 245, 259

Nasturtium (*Tropaeolum majus*) 29, 30, 31, **138–39**
   used in recipes 201, 235, 247
Nausea 206, 243, 244, **271**
Nettle (*Urtica dioica*) **140–41**
   used in recipes 218, 222, 223, 225, 234, 237, 238, 243, 247, 255
Neuralgia 241, 242, **289**
Nitrogen 30–1, 47, 86
No-dig gardening 19
Nosebleeds 248, **283**

Oils 183, 184, 185
   essential 185, 189, 194
   filters 186
   infusions 192–93
   paraffin oil 188
   preserving in 38
Ointments 183, 192, 193
   see *also* Herbal Recipes Index
Olive leaf extract 185
Oregano (*Origanum vulgare*) 15, 23, 27, 31, 36, **128–29**
   used in recipes 190, 202, 210, 234, 238, 240, 257, 259
Organic fertiliser 22, 32, 34
   recipe for 258
Organic herb gardening 17–18
Organic matter, and soil 17, 18, 19, 27
Organic potting soil mix 32
Osteo-arthritis 238, 239, 240, **274**
   see *also* Arthritis
Osteoporosis 274
Our Lady's thistle see Milk thistle
Overfeeding herbs 27

Pain 238, 239, 240, 241, 242, 254, **290**
Pansies see Heartsease
Parsley (*Petroselinum crispum*) 27, 30, **142–43**
   used in recipes 201, 202, 203, 205, 207, 208, 218, 223, 249, 254
Pathways 3, 15, 16, 19
Pelargonium (*Pelargonium* species) 23, 29, 30, **144–45**
   used in recipes 190, 213, 214, 215, 217, 219, 226, 228, 231, 246, 247, 252
Pennywort see Gotu kola
Peppermint see Mint
   used in recipes 190, 201, 221, 222, 223, 224, 225, 230, 231, 232, 233, 238, 244, 250, 257
Perennial herbs 21, 23, 27, 34,
Pests 29, 31
   aphids 29, 31, 110, 119, 139, 141, 157
   ants 99, 133, 163
   cabbage moth 31, 145
   caterpillars, leaf-eating 104
   cockroaches 133
   cutworm 29
   flies 55, 133, 151, 163, 215, 255
   mosquitoes 133, 151, 219
   pest-repelling herbs 29, 44, 45, 55, 56, 57, 59, 65, 70, 83, 99, 104, 119, 122, 123, 124, 133, 139, 145, 149, 151, 153, 163, 169, 175
   pest-repelling recipes 141, 157, 213, 215, 254, 255
   red spider mite 31
   snails and slugs 29, 45, 104
   tomato hornworms 29
   trap herbs 29, 139
   whitefly 119, 139
Phosphorous 80, 86, 141, 171
Pimples and acne 184, 218, 219, **264**
Placebo effect 179
Pollinators, attracting 29, 30, 89, 107, 140, 172
   see *also* Insects, beneficial
Poor man's saffron see Calendula
Potager (kitchen garden) design 13
Potassium 31, 75, 82, 86, 93, 131, 163, 177
Poultices
   herbal 193
   herbal honey 190
Predators 29, 89
   see *also* Insects, beneficial
Preparations, herbal **188–96**
Preservatives, natural 185, 196
Preserving herbs 36, 51, 54, 109, 121, 146, 165
   drying 36
   freezing 38
   in honey 38, 189–90
   in oil 38, 192
   in salt or sugar 36
   in vinegar 195–96
Propagating herbs 21–5
Prostate **277**
Psoriasis 183, 184, 251, 252, **264–65**
Pyrethrum 29, 30

Rashes 251, 252, 256, **265**
Raspberry (*Rubus idaeus*) **62–3**, 227, 228, 247, 249
Red spider mite see Pests
Rheumatism 238, 239, 240, **272–73**
Rocket (*Eruca vesicaria*) 15, 27, 30, **146–47**
   used in recipes 201
Rodent repellent 119, 163
Rooting powder 25
Rosemary (*Rosemarinus officinalis*) 13, 15, 21, 23, 25, 27, 29, 30, 31, 36, **148–49**
   used in recipes 192, 205, 206, 210, 212, 213, 214, 218, 222, 223, 228, 229, 236, 238, 239, 246, 248, 252, 254, 257
Rue (*Ruta graveolens*) 55, **150–51**, 239, 242

Safety, see Medicinal herbs
Sage (*Salvia officinalis*) 23, 25, 29, 30, 31, **152–53**
   used in recipes 190, 225, 232, 233, 235, 245, 246, 249, 257
Salad burnet (*Sanguisorba minor*) **154**
   used in recipes 248
Savory 36, **155**
Scabies **265**
Scarring (reduction) 251, 252, **265**
Scented garden theme 21
Sciatica 238, 239, 240, 241, 242, **274**
Seeds 21–3
Selecting plants 19, 21
Selfheal (*Prunella vulgaris*) **156**
   used in recipes 228, 243, 246, 248, 249,
Shakespearean garden theme 21
Shamanic healers 179
Shea butter 184
Shingles 238, 241, 242, **289**
Shock **279**
Sinusitis 194, 196, **283**
Site of garden 15, 17
Skin **260–67**
   dry and itchy 226, 227, 228, 251, 252, **261**
   itchy, irritated or infected 250, 251, 252, **263–64**
Snails and slugs see Pests
Soap, natural 184, 222, 223
Soapwort (*Saponaria officinalis*) **157**
   used in recipes 214, 222, 229
Soil, healthy 17–18, 19, 27, 30, 31, 32
Sore throat 230, 232, 233, 235, 236, 238, **283–84**
Sorrel (*Rumex acetosa*) **158–59**
   used in recipes 201, 202
Sourcing herbs 178, 186
Spiders see Insects, beneficial
Spiral garden design 13
Sprains see Strains and sprains
Sprays 31, 78, 79, 99, 105, 117, 133, 141, 157, 169
   recipes 193–94, 212, 213, 238, 255, 258, 259
St John's wort (*Hypericum perforatum*) **160–61**
   used in recipes 239, 241, 242, 245, 247, 249, 250, 256
St Mary's thistle see Milk thistle
Steam inhalation 194, 196
Stepping stone herb garden 15
Stevia (*Stevia rebaudiana*) **162**, 190
   used in recipes 220, 225, 247
Stings 184, 251, **266–67**
Stomach cramps 242, 243, **271–72**
Strains and sprains 238, 239, 240, **274**
Stress 246, 256, **277–78**
Summer savory (*Satureja hortensis*) see Savory
Sunburn (and burns) 183, 190, 196, 251, **260, 265–66**
Sweets 194
   recipes for 195, 232, 233, 235, 244
Syrups 194–95, 235
   recipes for 194, 195, 204, 235, 237

Tansy (*Tanacetum vulgare*) 29, 30, **163**
   used in recipes 215, 239, 254, 255, 257, 259
Tarragon (*Artemisia dracunculus*) **164–65**, 181
Tea tree (*Melaleuca alternifolia*) **166–67**
   essential oil 185, 194
   used in recipes 212, 213, 218, 219, 221, 224, 238, 251, 255, 259

Tendonitis 274
Tension 246, 249, 256, **277–78**
Terminators see Insects, beneficial
Themes for gardens 21
Throat see Sore throat
Thrush **287**
Thyme (*Thymus* species) 15, 21, 23, 25, 27, 30, 31, 36, **168–69**
   creeping, as ground cover 27
   essential oil 185
   used in recipes 190, 206, 207, 208, 210, 212, 230, 231, 232, 233, 234, 235, 237, 238, 243, 247, 257, 259
Thymol 59, 129, 169
Tinctures 195
Tonic
   general 237, **285–86**
   uterine **288**
Tonsillitis 232, 233, 235, 236, 238, **284**
Tools and equipment 186
Toothache and teeth 225, 241, **284**
Trap herbs 29, 139
Trimming herbs 17, 27, 34

Urinary ailments 245, **277**
Using herbs with intent 178–79
Uterine tonic **288**

Varicose veins 193, 247, **266**
Vegetable gardening see Companions, herbal
Vermicompost (earthworm castings) 22
Vermiculite 22, 23, 32
Vietnamese coriander (*Persicaria odorata*) **170**
Vietnamese mint see Vietnamese coriander
Vinegars 194, **195–96**
   recipes for 200, 236
Viola see Heartsease
Vitamin E oil 185
Vomiting **271**

Wasps see Insects, beneficial
Watercress (*Nasturtium officinale*) **171–72**
   used in recipes 201, 202, 218, 225, 247
Watering 27
   seeds 22, 23
Weeds 27
Whitefly see Pests
White willow tree 177
Winter savory (*Satureja montana*) see Savory
Wild garlic (*Tulbaghia violacea*) **173**
Witch hazel 184, 189
Woman's herb see Raspberry
Worms 17
   cutworms 29
   earthworms 17, 18, 92
   leaf-eating 45
   tomato hornworms 29
Wounds 191, 251, 252, 256, **266–67**
   see also Blood vessels and bleeding
Wrinkles 216, **260**

Yarrow (*Achillea millefolium*) 30, 31, 32, **174–75**, 179
   used in recipes 231, 234, 243, 248, 249, 259

HERBAL RECIPES INDEX

Beauty care 216–29
   Anti-bug balm 219
   Anti-funky foot balm 224
   Conditioner for dry hair 223
   Cooling eye gel 228
   Edible body balm 220
   Evening primrose face cream 227
   Exfoliating body scrub 221
   Frankincense and rose skin cream 219
   Gardener's hand cream 228
   Gardener's hand scrub 229
   Hair-strengthening juice 218
   Lavender and neroli body powder 221
   Lemon fresh rose petal bath bombs 226
   Loofah scrubber soap 222
   Perfume oil or balm 217
   Pimple fixer 218
   Puffy eye treatment 227
   Rosemary and peppermint shampoo soap 223
   Rose-scented pelargonium hand cream 226
   Sage and peppermint toothpaste 225
   Shiny shampoo 222
   Tinted lip balm for dry lips 224
   Toner for oily skin 219
   Wrinkle and crinkle balm 216
   Yummy mummy body butter 220

Common ailments 230–53
Aches, pains, sprains and bruises **238–42**
   Achy painy anti-inflammatory tea 238
   Headache balm 239
   Hot diggedy joint salve 240
   Massage salve for aches and pains 239
   Neuralgia brew 241
   Shingles soother 242
Coughs, colds, sore throats, flu and hay fever **230–38**
   Angelica syrup and cough sticks 235
   Chest rubs 231
   Cold and flu tea 230
   Echinacea cough lozenges 232
   Elderberry and sage cough syrup 235
   Frequent flyer balm 236
   Hay fever G & tea 234
   Honey and cinnamon cough sweets 233
   Immune-boosting tonic 237
   Onion, ginger and nasturtium syrup 235
   Plum blossom vinegar 236
   T & T tonsil spray 238
   Tea thyme 230
Digestion, stomach and urinary system **242–45**
   After-dinner mint liqueur 243
   Hair-of-the-dog detoxifying tincture 243
   Horseradish and cranberry bladder cleanser 245
   Indigestion tea 242
   Tense tummy tincture 243
   Travel-sickness sweets 244
Heart, blood and circulation **247–48**
   Anti-anaemia tea 247
   Cellulite body butter 248
   Hibiscus healthy blood tea 247
   Nosebleed balm 248
Skin conditions **250–52**
   Cold-sore balm 250
   Green goodness ointment 251
   Honey-and-herb wound healer 252
   Skin-soothing cream 252
   Sunburn soother 251
Tension, stress, mind and emotion **245–47**
   Brain-boosting tonic wine 246
   Dr Feelgood tea 247
   Dream enhancer 245
   Pick-me-up balm 246
   Sleep-easy tea 245
Women 249
   Menopause tea 249
   Time-of-the-month brew 249

Garden, in the 258–59
   Anti-fungal spray 259
   Comfrey tea 258
   Insect-repelling spray 259

House, around the 212–15
   Air-freshener 213
   All-purpose cleaner 212
   Fridge freshener 214
   Insect repellent 215
   Moth-repellent mix 213
   Mould cleanser 213
   Scouring paste 214
   Soapwort washing liquid 214

Pets, just for 254–57
   All-purpose healing balm 256
   Arthritic relief for older dogs 254
   Catnip mouse 257
   Fly spray for dogs and horses 255
   Healthy coat tonic 255
   Herbal flea powder 254
   Herbs for hens 257
   Stress and tension easer 256

Tasty temptations 200–11
   Elderflower syrup 204
   Elderflower vinegar 200
   Green herb soup 202
   Hangover kicker 206
   Herb and honey purée 207
   Herb butter 207
   Herb oil 208
   Iced herb bowl 209
   Kiwi fruit iced lollies 208
   Lemon and mint zinger 204
   Mediterranean salad 203
   Plum, chocolate and mint sauce 211
   Rosemary and lime salt 206
   Spring booster salad 201
   Therein lies the rub 210
   Winter warmer soup 205

HERBS PICTURED ON THE FOLLOWING PAGES: endpapers, Sage; page 2, Russian sage; page 4, White echinacea; page 6, Spring onion buds; and page 10, Lemon thyme.

SUNBIRD PUBLISHERS

First Published in 2012

Sunbird Publishers (Pty) Ltd
The illustrated imprint of Jonathan Ball Publishers (Pty) Ltd
P O Box 6836
Roggebaai 8012
Cape Town, South Africa

www.sunbirdpublishers.co.za

Registration number: 1984/003543/07

Copyright published edition © Sunbird Publishers (Pty) Ltd
Copyright text © Jane Griffiths
Copyright images © Jane Griffiths
Copyright images © on pages 58, 68, 110 and 127 Zirkia Swart, Mountain Herb Estate
Copyright images © on pages 126 and 130 iStock
Copyright images © on page 60 Random Harvest Indigenous Nursery

www.janesdeliciousgarden.com

Design, typesetting and cover design by MR Design
Project management by Michelle Marlin
Editing by Kathleen Sutton
Proofreading by Patricia Myers Smith
Photographs by Keith Knowlton and Jane Griffiths

Reproduction by Resolution Colour (Pty) Ltd, Cape Town
Printed and bound by Tien Wah Press (Pte) Ltd, Singapore

Jane Griffiths asserts the moral right to be identified as the author of this work. All rights reserved. No part of this publication may be reproduced, stored in a retrieval system or transmitted, in any form or by any means, electronic, mechanical, photocopying, recording or otherwise, without the prior written permission of the copyright owner(s).

ISBN 978-1-920289-54-6

While every last effort has been made to check that information in this guide is correct at the time of going to press, the publisher, author and their agents will not be held liable for any damages incurred through any inaccuracies.

Disclaimer

Please note that any information published in this book is done so solely for educational purposes. It does not constitute any medical advice whatsoever, and nor does it replace medical attention or diagnosis. Consult a qualified healthcare practitioner for the diagnosis and treatment of any disease, ailment or medical condition.